Reflexive biotechnology development

Reflexive biotechnology development

Studying plant breeding technologies and genomics for agriculture in the developing world

Wietse Vroom

Wageningen Academic Publishers

ISBN 978-90-8686-106-4

Cover design and photos:
Wietse Vroom

First published, 2009

Table of contents

Acknowledgements **11**

Acronyms and abbreviations **15**

Chapter 1
Introduction: genetic technologies for international agricultural development **17**
The era of development 17
Agricultural modernisation for development 21
Genetic technologies for agriculture 28
From Green Revolution to Gene Revolution? 34
Making genetic technologies 'appropriate' for agricultural development 37
Concluding remarks 39
Structure of the thesis 42

Chapter 2
Research design **45**
Motivation and objectives of the study 45
Research questions 46
Approach – explorative qualitative research 47
Case study selection 47
Data collection – a technographic approach 51
Validity of the study 55
Concluding remarks 57

Chapter 3
Biotechnologies and the transformation of agricultural production systems **59**
Introduction – Historical trends in agricultural modernisation and industrialisation 59
Agricultural modernisation – From an imposed condition to an open-ended approach 61
Industrialisation of agriculture – Issues of control in the Third Agro-Food order 65
Intellectual property – The crucial importance of ownership 72
Recapitulation 78
The relationship between technical design and social structures 81
In conclusion: key elements of a conceptual framework 88

Chapter 4
Transgenic insect resistance for the poor – Towards a win-win-win situation in Indian vegetable farming **95**
Introduction 95
The setup of a public private partnership for the development of Bt Brassica 96
The material reconstruction of transgenic technology 100
Tailoring the mode of commercialization 103
Representation of farmers, rather than participation 106
The role of intellectual property and liability 109
Discussion – Farmers as recipients of technology 113

Chapter 5
Reconsidering the role of potato farmers in breeding and multiplication – Experiences of the International Potato Centre **117**
Introduction 117
The International Potato Centre – Producing global public goods for potato farmers 119
Challenging a trend towards genetic erosion in potato cultivation in the Andes 123
Enabling farmers' seed potato production 131
Working with cultural connotations – Genetic fingerprints as Kipu diagram 135
Discussion – Breeding technologies for a non-industrializing development strategy 137

Chapter 6
Linking upstream genomics research with downstream development objectives – The challenge of the Generation Challenge Programme **141**
Introduction 141
The Generation Challenge Programme – Upstream genomics research for pro-poor agricultural innovation 143
Targeting the poor – Farming systems, crops and traits 146
Challenges for a science-led research programme 148
Complementary innovation systems 151
Complementarity in practice: different research partners and technology as a service 154
Discussion – The potential for a service-like approach to agro-technological innovation 158

Chapter 7
Discussing the diversity in approaches to agro-technological innovation **161**
Introduction 161
Comparative analysis of the cases – Multiple dimensions of appropriateness 167
Additional reflections on the openness of innovation 177
Flexibility in the relationship between technological design and social meaning 180
Implications for innovation policy and questions for future research 183
In conclusion – Reflexive biotechnology development 189

References **193**

Summary **209**

About the author **215**

Acknowledgements

This thesis wouldn't have been possible without the help and support of a great number of people. First of all, I'd like to thank NWO-MCG for their funding which has made this PhD project possible (Dossier no. MCG-CO3-02), and for their additional travel grant which allowed me to participate in the 'Reconstructing Agro-Biotechnologies for Development?' conference in Kyoto, November 2007 (Dossier no. 050-33-043). In addition, I'd like to thank WTMC research school and the Faculty for Earth and Life Sciences of the Vrije Universiteit Amsterdam for the financial support in the printing costs of this thesis.

On a more personal note, there are a lot of colleagues and supervisors to thank. Hardly having finished my Masters in Molecular Sciences with a pinch of Communication Sciences, I was lucky enough to meet Guido Ruivenkamp and Joost Jongerden. They didn't mind my somewhat primitive interest in technology and sociology and were kind enough to take me on board as a PhD candidate on a new programme, investigating the potential for genomics and biotechnology in developing world agriculture. This leap into the dark turned out to be a very stimulating experience, and it got me in touch with – to me – entirely new academic worlds of 'Science and Technology Studies', and 'Development Studies'. I was surprised, amazed, and sometimes puzzled about the questions being asked, assumptions being made and methodologies being followed. I did enjoy it, but whether all of that was ever going to amount to a coherent and remotely useful PhD thesis, I wasn't quite sure of (and whether it has, I'd rather leave up to the interested reader). I am very grateful for the trust that Guido and Joost gave me in those somewhat insecure and exploratory phases of the project. Although intellectually demanding supervisors, throughout the project they have given me a lot of intellectual and practical freedom in shaping this research. This even led to their approval of my plans to follow my wife Marleen to Guatemala, where she'd started a development project. I've spend the major part of my final year writing my thesis at a significant distance from Wageningen. This may not always have made our communication on new drafts easier or more efficient, but the fact that we succeeded in bringing this project to a good end bears witness to the productive, constructive and very pleasant working relation we had built over the first three years of my project.

I am also very grateful for the support of Steve Hughes and Joske Bunders who helped setting up this project and acted as co-supervisors (in the case of Joske only in the initial phases). Having a supervision team of four members may be quite a practical challenge, but it did present an incredible richness in information, expertise, and approaches that I and my fellow PhD candidates could more or less pick from at will. Over the years, the collaboration has been more and less intense in phases, but I did highly appreciate the interactions with Joske in Amsterdam, and with Steve in Exeter and on numerous other occasions.

Eric Deibel and Daniel Puente have been wonderful travel companions on this scientific journey. We simultaneously started our PhDs within the same programme and although that didn't necessarily mean we worked together on a common project, it did give me a strong sense of 'being in this together'. Although we clearly had very different backgrounds, styles and approaches to the scientific enterprise, I think we did manage in finding a common language in which we had a lot of valuable interaction. The least I can say is that I greatly enjoyed our companionship in this project and the moments of laid-back philosophizing, chatting and discussion. I'm extremely grateful for the comments and support given by you both, which has often helped me of a dead-end track, or which inspired me to venture into new directions.

Then there is an endless number of other colleagues and friends whose presence, support and comments I have greatly enjoyed. First of all, I want to mention my fellow PhD students and lecturers from the WTMC workshops and Summer Schools. These moments must have been among the most intellectually stimulating and enjoyable in my PhD life. A special thanks goes out to Sally Wyatt and Els Rommes for being such wonderful scientific mothers to all of us. I also want to extend my thanks to the members of the reading committee of this thesis, who have given me some very valuable comments and the inspiration to go the extra mile in the last stages of my research. Then there are the colleagues from the Athena Institute, the Corsage crew, the somewhat confused but greatly enjoyable OPNA gang, the great people from CSG, and the colleagues from the Applied Philosophy group with whom we share a corridor. There's too many of you to mention individually, but I do want to mention Bram de Jonge, who hasn't only been a great intellectual sparring partner and a great co-organiser of a conference on Intellectual Property in April 2008, but who has also become a warm friend. Having mentioned the April 2008 workshop, I also want to express my thanks to Niels Louwaars who has been a wonderful 'partner in crime' and 'mentor', and Frans van Dam who has opened so many doors for us at CSG headquarters. Finally, Inge. Where would I have been without our daily shot of caffeine and gossip about life at the Leeuwenborch, life in general and all else that matters in those 15 minutes of strictly non-scientific chat?

I feel privileged to have been in the position to travel a lot for my research, and to interview a great number of highly interesting people. Travels to India, Peru and Mexico have been defining experiences in both my professional as well as in my personal life. Interviewing scientists, research managers, policy makers, NGO members and countless others has been an extremely valuable experience. I have gained a lot of sympathy for your hard work and commitment to making the world a better place, and I've often had the wish that I could contribute more. My research has invited me to be a critical observer, but I hope I have been able to do your work justice. And although my research questions were different than yours, I hope I have been able to provide interesting and useful new perspectives on issues of agricultural and technological development.

Then there's family and friends, whose support is invaluable not so much because of their scientific input, but because of helping you out in any other way possible. There's a countless number of friends that have had a drink with me while I was letting of steam after a hard day at the office. Thanks for pulling me through. Also my parents have been an endless source of support, helping me to reflect upon my expectations of this project, my own role as a PhD student, and the ways I could keep a grip on whatever I was doing. Thanks for your sincere interest in my work, and your support in seeing the bigger picture. Finally, Marleen. Your love makes any scientific problem seem so insignificant that it makes all work lighter. And by making it lighter, you've made it more playful, enjoyable and inspired than I ever thought possible. Thank you for being there. I'll always love you.

Acronyms and abbreviations

AATF	African Agricultural Technology Foundation
ABSP	Agricultural Biotechnology Support Programme
AKIS	Agricultural Knowledge and Information Systems
AVRDC	Asian Vegetable Research and Development Center (The World Vegetable Centre)
Bt	*Bacillus thuringiensis*
CBD	Convention on BioDiversity
CESAR	Centre for Environmental Stress and Adaptation Research
CGIAR	Consultative Group on International Agricultural Research
CIMBAA	Collaboration on Insect Management for Brassicas in Asia and Africa
CIMMYT	Centro Internacional de Mejoramiento de Maiz Y Trigo (International wheat and maize improvement centre)
CIP	Centro Internacional de la Papa (International Potato Centre)
CMS	Cytoplasmic Male Sterile
CP	Challenge Programme
DBM	DiamondBack Moth
DNA	DeoxyriboNucleic Acid
DUS	Distinct, Uniform and Stable
EMBRAPA	Empresa Brasileira de Pesquisa Agropecuária (Brazilian Agricultural Research Corporation)
Ex situ	Latin for 'off site'. In the context of '*ex-situ* conservation', it refers to the conservation of traditional varieties in seed banks, or otherwise outside of their natural habitat, in contrast to '*in-situ* conservation'
FAO	Food and Agriculture Organisation
FTO	Freedom To Operate
GCP	Generation Challenge Programme
GM	Genetic Modification
GMO	Genetically Modified Organism
GSS	Genotyping Support Service
HFCS	High Fructose Corn Syrup
HYV	High Yielding Variety
Ibid.	*Ibidem*, Latin for 'in the same place'. In the text this term refers to a repeated reference to a previously mentioned source.
In situ	Latin for 'in the place', or 'on site'. In the context of '*in-situ* conservation', it refers to the conservation of traditional varieties through their ongoing cultivation in their natural habitat, in contrast to '*ex-situ* conservation'.

INIA	Instituto Nacional de Investigación Agraria (National Institute for Agricultural Research)
INIFAP	Instituto Nacional de Investigaciones Forestales, Agrícolas y Pecuarias (National Research Institute for Forestry, Agriculture and Animal Husbandry)
IP	Intellectual Property
IPM	Integrated Pest Management
IRRI	International Rice Research Institute
ISAAA	International Service for the Acquisition of Agri-biotech Applications
LMO	Living Modified Organism
Ltd.	Limited
MDG	Millennium Development Goal
MMB	Monsanto-Mahyco Biotechnology
NARS	National Agricultural Research System
NGO	Non-Governmental Organisation
NRI	National Resources Institute
OECD	Organisation for Economic Co-operation and Development
OPV	Open-Pollinated Variety
PBR	Plant Breeders Rights
PGSC	Potato Genome Sequencing Consortium
PhD	*Philosophiæ Doctor* (Research Doctorate)
PIPRA	Public sector Intellectual Property Resource for Agriculture
PRA	Participatory Rural Appraisal
PROINPA	Promoción y Investigación de Productos Andinos (Promotion of and Research on Andean Products)
PVP	Plant Variety Protection
Pvt.	Private
QTL	Quantitative Trait Locus
R&D	Research and Development
RRA	Rapid Rural Appraisal
SP	Sub-Programme
SSR	Single Sequence Repeats
STS	Science and Technology Studies
TILLING	Targeting Induced Local Lesions IN Genomes
TRIPS	Trade Related Aspects of Intellectual Property Rights
UK	United Kingdom
UN	United Nations
UPCH	Universidad Peruana Cayetano Heredia
UPOV	Union pour la Protection des Obtentions Végétales (The International Union for the Protection of New Varieties of Plants)
US	United States
USAID	United States Agency for International Development

Chapter 1
Introduction: genetic technologies for international agricultural development

"We can realistically envision a world without extreme poverty by the year 2015 because technological progress enables us to meet basic human needs on a global scale and to achieve a margin above basic needs unprecedented in history."
(Jeffrey D. Sachs 2005, p. 347; Director of the Millennium Development Project)

"The idea of development stands as a ruin in the intellectual landscape. Delusion and disappointment, failures and crime have been the steady companions of development and they tell a common story: it did not work."
(Wolfgang Sachs 1992, p. 1)

"No period in history has been more penetrated by and more dependent on the natural sciences than the twentieth century. Yet no period, since Galileo's recantation, has been less at ease with it."
(Eric Hobsbawm 1995, p. 522)

The era of development

We are living in an age of Millennium Development Goals; a set of eight, time-bound and measurable objectives to eradicate global hunger and poverty before 2015. Any contemporary project aiming at international development takes place against the background of this international ambition to do something about global inequality. Pleas have been made to solve problems of under- and malnutrition, find a cure for some of the most devastating diseases plaguing humanity, to increase the availability of clean water, to increase levels of education and to combat environmental degradation; all of this especially in the poorer regions of this world (Box 1.1). Whether these goals will actually be met in 2015 remains highly questionable at the time of writing this thesis.[1]

[1] The most recent 2007 'Millennium Development Goals Report' remains optimistic about the possibilities of still reaching all MDGs by 2015, but admits that success so far has been "uneven" (United Nations 2007).

Box 1.1. The eight UN Millennium Development Goals.

1. Eradicate extreme poverty and hunger
2. Achieve universal primary education
3. Promote gender equality and empower women
4. Reduce child mortality
5. Improve maternal health
6. Combat HIV/AIDS, malaria and other diseases
7. Ensure environmental sustainability
8. Develop a global partnership for development

Source: www.un.org/millenniumgoals (last accessed 17 September 2008).

The Millennium Development Goals (MDGs) have been put forward at the 2000 United Nations Millennium Summit.[2] They are accompanied by a Millennium Declaration which includes a wide range of commitments to human rights, good governance and democracy. While presented as a new initiative, in fact the MDGs are the successors to similar and earlier formulated development goals at the 1995 Copenhagen UN World Summit on Social Development, and a set of development goals agreed upon by the World Bank and OECD countries (Thomas 2000, p. 3-4).[3] Their content is the outcome of decades of international debate, research and activism, in which many independent organisations have left their marks on the formal international development agenda.

These recent declarations of international development goals reflect a desire to approach underdevelopment in a globally coordinated way, thereby increasing the impact of development programmes. In fact, the adoption of the Millennium Development Goals is the most recent climax in what has been called the 'era of development' (Thomas 2000, p. 5). These goals have become iconic of contemporary well intended efforts to do something about global inequality; of the efforts to bring global food production and health care from the shadows of underdevelopment into the light of modernity. As such, this set of goals illustrates contemporary ideas that international development is not only desirable, but also achievable given the right amount of investments, and given the right strategy. They are landmarks of a specific discourse and ideology on international development; one in which the use of modern technologies gains an important place and function.

[2] See http://www.un.org/millenniumgoals (last accessed 17 September 2008).

[3] OECD = Organisation for Economic Co-operation and Development; with membership of 30 developed and industrialized countries.

This PhD thesis is concerned with the question how technologies are being used for international development, and how that process of technological innovation is interrelated with social change, and with implicit assumptions about 'progress'. More specifically, this thesis zooms in on agricultural development, and within that sector on the use of genetic technologies in plant breeding for farmers in developing countries. Against the background of persistent poverty and hunger in many parts of the world, it is hardly surprising that there is a strong call for the modernisation of agriculture, facilitated by the introduction of new technologies. However, the way in which modern technologies are harnessed in order to improve agricultural production does raise questions regarding the relationship between technological development, and the existing social order. Choices of how food is produced in the future go far beyond mere technical or economic considerations alone. They involve important questions regarding the role of farmers in agricultural innovation and production systems. But who gets to answer these questions? And how are the answers to these questions reflected in the methodologies and technologies of agricultural development projects?

Reflexive development

Questioning development is not new. In spite of a general agreement on the need to address global inequality, both the process and the ends of development have been heavily debated. A crucial question that arises is whether development – aiming at a greater quality of life and a more just distribution of wealth – requires or implies a process of modernisation in which production and trade are rationalized according to a Eurocentric model of industrialisation.[4] Critical comments on agricultural and economic modernisation have made clear that it is not always a smooth ride into modernity, but instead a heavily contested and sometimes painful process of social transformation. Concerns about sustainable development, an erosion of identity and culture, and unequal differentiation of development benefits are characterising the contemporary development debate as much as its projected benefits for the global community. This has led to pleas for 'Alternative Development', which generally agree upon the need for development, but argue for a more participatory, democratic, people-centred development process (Hettne 1990; Max-Neef 1991). More radical are notions of 'Post-Development', which do not seek 'alternative development', but 'alternatives *to* development', arguing that the notion

[4] Strict definitions for both 'development' and 'modernisation' are hard to find, and a full exposé on different interpretations would go beyond the scope of this chapter or thesis. The relevant difference between both terms in the ways they are used here is that development (as the quest for a better quality of life) can be defined locally and in very different ways, and is not restricted to economic parameters. Modernisation in contrast is generally associated with a rationalization of production and trade and is in that sense biased towards a dominant model of Eurocentric industrialisation. This narrow interpretation of the term modernisation is both countered by and confirmed by a quest for 'alternative modernities', in which a process of modernisation no longer signifies a conversion to a single modernity, but allows for locally defined and divergent 'modernities'. See Gaonkar (2001b) and Taylor (2001) for a discussion of different perspectives on modernity.

of development itself is fundamentally flawed and essentially leads to a 'Westernalization' of the world (Sachs 1992; Latouche 1993; Escobar 1995).

The differences between mainstream-, alternative-, and post-development concepts are important, but arguably more interesting is the observation that mainstream development has changed over the years, and has taken on board elements that once belonged to the alternative development discourse. For example, Jan Nederveen Pieterse argues that the commitment to values of participation, sustainability, and equity is being widely shared, not only among non-governmental organisations (NGOs), but also in the world of UN agencies including the World Bank (Nederveen Pieterse 1998). Moreover, when mainstream development is simplified as a single, homogeneous thrust toward modernisation, its diversity, complexity and adaptability are often underestimated. Therefore, rather than continuing the false dichotomy between mainstream-, and alternative- or post-development, he argues for a more fruitful position of 'reflexive development'. This notion shifts focus to the ways in which development policy increasingly becomes concerned with the management of development interventions itself, and takes on board some of the criticisms that are levelled at it (Nederveen Pieterse 1998).[5]

This reflexivity of development processes and policies is a useful entry point to start questioning contemporary approaches to international agricultural development, and the ways in which they harness agro-biotechnologies to improve agricultural production. Apart from the fact that the notion of 'reflexive development' takes the analysis away from a polarized comparison of conventional agricultural development and alternative approaches, it raises new questions. Most importantly, it raises the question in what ways exactly contemporary development projects are taking criticisms and concerns on board in their work, and how that influences the way in which they design and apply new technologies for the sake of agricultural development. Reflexivity is a useful term to capture the flexibility, adaptability and versatility in (technology) development approaches. However, at the same time important differences may be witnessed in terms of the nature and extent of reflexivity in different projects. While some values in development – like participation or sustainability – are widely shared, it does not mean that they are widely and evenly practiced, or operationalized in the same way.

[5] Nederveen Pieterse presents the notion of 'reflexive development' as a corollary to 'reflexive modernisation' as famously described by Ulrich Beck. Beck contrasts 'simple modernity' concerned with 'mastering nature' with reflexive modernity, the condition in which the moderns are increasingly concerned with managing the problems created by modernity itself (Beck 1992). Nederveen Pieterse indicates parallels in the way modernity and 'progress' are questioned in reflexive modernity and -development. For example, he mentions the breakdown of faith that technical progress equals social progress, which is typical for reflexive modernity. This is matched by a parallel questioning in reflexive development: does growth equal development, and does economic growth equal social development (Nederveen Pieterse 1998)? He argues that such questioning of modernity or development is no longer external to a mainstream discourse, but inherently part of the dynamics of reflexive modernity/development.

Decades of critical studies of technology, and 'science and technology studies' (STS), have stressed the intricate relationships between technological development, social structures and power relations, and have argued that technological development cannot be understood in technical terms alone, but requires an analysis of the social relations in which technologies are developed and applied (MacKenzie and Wajcman 1985; Ruivenkamp 1989; Bijker 1995).[6] Against this background, it seems reasonable to assume that the ways in which contemporary projects of agricultural development respond to tensions in their work, will be influenced by the socio-political and institutional context in which they take place. Any project may be expected to be sensitive of some of the controversies in international development, especially in the heavily contested terrain of agro-biotechnology development. For the same reason, any project may be expected to have found ways to respond to and deal with the controversies in technology development, and the challenges of making technology work for agricultural development. But the important point open for investigation is whether this leads to anything more than a superficial, instrumental adaptation of development projects to the most controversial issues in public debate. To what extent are contemporary projects of agricultural development reflexive in their approaches, and can they meaningfully challenge not only the technological means to agricultural development, but also the kind of modernity they are contributing to?

Chapters 1 and 3 will unpack and elaborate this critical perspective upon 'reflexive development' in the context of agro-biotechnologies for international agricultural development. The first chapter will focus on the significance of agricultural development, and the importance of new genetic technologies in this development. Most importantly, it will introduce a notion of 'appropriateness' of biotechnologies for agricultural development and start the discussion on how to conceptualize this notion. Chapter 2 is dedicated to the research design of this study and introduces the main research questions and methods of data collection. Then, Chapter 3 will elaborate on a historical context, with processes of modernisation and industrialisation in which agricultural development is taking place. It will also review and discuss different conceptualizations of technologies, outlining the relationship between technical design and the wider social order. These elements lead to a sharpening and elaboration of the main research questions presented in Chapter 2, and provide a starting point for the analysis of several case studies of agro-biotechnology development for international agricultural development in the later chapters.

Agricultural modernisation for development

The link between international development and agriculture is not coincidental or arbitrary, considering that agriculture is widely acknowledged to play a key role in the economic

[6] A further elaboration of the significance of science and technology studies and critical studies of technology, will be undertaken in Chapter 3.

development of less developed countries (Thirtle *et al.* 2001; Dorward *et al.* 2004; Diao *et al.* 2007). Moreover, it has a very direct link with the availability of sufficient quantities of good quality food, which is an important precondition for food security.[7] As a result, a significant part of the development debate focuses on the improvement of agricultural production, through a wide range of potential interventions. Depending on the main problems in an agricultural production system, development may focus on productivity increases, diversification of production, reducing the costs and risks of cultivation, or making food production more sustainable by reducing environmental impacts. This means that a wide range of interventions and tools are being used, ranging from the introduction of irrigation, fertilizers, improved pest management strategies, new crop varieties, improved post-harvest conservation methods, and even improved access to markets. Within this wide range of potential strategies and entry points to agricultural development, the potential of improved crop varieties is an area that receives significant attention in the international debates on agricultural development, and that is one of the main activities of the Consultative Group on International Agricultural Research (CGIAR).

Several reasons might be indicated that could legitimate a special interest for new crop varieties, and hence for plant breeding. One is that breeding is an attractive entry point for international contributions to local agricultural production. Improved varieties may be useful in a wide range of circumstances, while much of the pre-breeding work can be done in isolation of the local situation. While the installation of irrigation facilities, or the provision of improved fertilizer requires a project to directly engage with a local situation and its complex dynamics, early phases in plant breeding generally allow for a much more distanced engagement with the problems in agricultural production. Commonly, only downstream variety development is conducted in close contact with farmers and within specific environmental conditions in which the improved variety is supposed to perform. This approach is reflected in the work and institutional organisation of a series of specialized plant breeding institutes of the CGIAR, that provide crop varieties for a very wide range of countries and regions, but have centralized their upstream pre-breeding research to an important extent in the international research centres. Plant breeding is a strategic investment in that sense, which can lead to potential benefits in a wide range of localities. Having said that, there are some pitfalls in centralizing (pre-)breeding, in the sense that crucial interactions between new crop varieties and local conditions may be different than predicted or expected. For that reason, increasing attention has been going out to variety development with locally adapted crop varieties, and to participatory methodologies to investigate local needs and priorities.

Secondly, seed plays a crucial role in agricultural production, and gathers a wider range of problems and potential solutions for agricultural production. While problems with productivity

[7] A precondition, but not sufficient, considering that hunger can prevail in the presence of abundance of food if those with the greatest need for food lack purchasing power (Sen 1981).

and pest infestation can be addressed by improved soil management, irrigation, fertilization, and improved pest management, they may also be addressed by plant breeding. Modern plant breeding is increasingly capable of producing plants which are resistant to diseases and pest insects, which are capable of growing under harsh environmental conditions, and which are increasingly productive because of a more efficient use of nutrients. All these aspects of agricultural production are gathered in the nature and quality of the seed, and breeding therefore provides a highly strategic way of engaging with agricultural production. This strategic role of the seed has of course not gone unnoticed by the plant breeding industry, which has been enthusiastic in claiming ownership on the seed through both legal and biological mechanisms (Kloppenburg 1988). This strategic aspect of the seed, not only in production but also in the political economy of plant breeding, provides the owner and developer of seed with a crucial and powerful role in the agricultural production system, as will be further discussed in Chapter 3.

Thirdly, the introduction of improved varieties has proven to be a highly effective way of influencing agricultural productivity during the Green Revolution in the second half of the twentieth century. In fact, current assumptions about the potential of agricultural modernisation for economic development, can in general be traced back to this extremely important experience in the planned, large scale modernisation of agriculture in developing countries. Given the crucial importance of the Green Revolution in our current understanding of agricultural modernisation, a brief review of this process is appropriate. It will provide an important background to contemporary discussions on the role of biotechnologies and new crop varieties in agricultural development, and the different perceptions of what agricultural modernisation is all about.

The controversy of the Green Revolution

'Green Revolution' is the name that was given to a process of agricultural modernisation in developing countries, most notably in the 1960s and 1970s.[8] It was aimed at the increase of agricultural productivity, and depended upon a combination of improvements in infrastructure and research capacity, and the transfer and introduction of relatively simple agricultural technologies. These novel agricultural technologies included modern high yielding varieties (HYVs) of rice and wheat, and a package of agricultural tools and practices, such as the use of chemical fertilizers, irrigation and pesticides.

Arguably the most interesting and innovative aspect of the Green Revolution was the development of 'miracle' dwarf varieties of wheat and rice, which had a shorter plant

[8] The term 'Green Revolution' was first used by USAID administrator William Gaud in a speech entitled "The Green Revolution: Accomplishments and Apprehensions" before the Society for International Development, on March 8, 1968.

morphology which allowed the crop to spend its energy on making grain, instead of stems or leafy material. This specific crop morphology allowed the new varieties to respond much better to the application of high quantities of chemical fertilizer, which led to strong yield increases (Khush 1999).[9] Under good conditions (with irrigation or plenty of rain, and fertilizer) the HYVs strongly outperformed traditional varieties of wheat and rice. Under more difficult (rainfed) circumstances, the advantage of modern varieties was generally less clear, and traditional varieties sometimes proved to be more productive (e.g. Negi 1994).

The start of the Green Revolution can be traced back to the invention of dwarf varieties of wheat by Norman Borlaug in 1954, at the research centre that is now known as CIMMYT (*Centro Internacional de Mejoramiento de Maiz Y Trigo*: International wheat and maize improvement centre) (Parayil 2003). Equally important was the later development of dwarf varieties of rice at the International Rice Research Institute (IRRI).[10] Govindan Parayil argues that the work of these international research institutes was seminal for the success of the Green Revolution, but that there were a number of other crucial protagonists in this process. These included local and national governments of developing countries, who increased their budgets for agricultural research, and planned and coordinated the transfer and adoption of new technologies through various national institutions. In addition, multilateral and bilateral donor agencies played an essential role in supporting the setup of agricultural universities according to the American model of land-grant universities (US Agency for International Development; USAID), in the development of national agricultural research systems (Rockefeller Foundation), and in farm extension work (Ford Foundation) (Parayil 2003).

The Green Revolution has been a success in terms of productivity increases in cereals, and adoption of the improved varieties by farmers, at least in some areas. Several studies provide productivity statistics that demonstrate that rice and wheat yields more than doubled within two decades, in countries like India, Pakistan, the Philippines, Mexico, Turkey and Indonesia (Conway 1998; Pingali and Heisy 1999). Also, Evenson and Gollin provide data suggesting that the development of modern cereal varieties has led to a prolonged increase in productivity, which in fact has had the greatest effect in the 1980s and 1990s (Evenson and Gollin 2003). They explain this effect by arguing that successive generations of new varieties have been developed, each contributing gains over previous generations.

[9] In addition to crop morphology, a number of other traits were modified as well, that increased the adaptability and yield stability of the new wheat and rice varieties. These included traits that allowed the crops to be planted at any time of the year and shortened the growth period, leading to increased cropping intensity. In addition, traits for disease and insect resistance were incorporated, as well as modest tolerance to soil salinity, alkalinity and metal toxicity (Khush 1999).

[10] The important role of these centres during the Green Revolution led to the instalment of a range of other international agricultural research centres, which in 1971 were brought together as the aforementioned Consultative Group on International Agricultural Research (CGIAR).

On the other hand, both the critics and proponents have noted that the benefits of the Green Revolution have been unevenly spread, for a variety of reasons. One crucial element has been that the Green Revolution explicitly focused on the uptake of modern farming practices by medium size and large scale farmers; leaving small scale farmers – and notably female farmers – in less favourable areas largely behind (Momsen 1991; Parayil 2003, p. 976). This allegedly exacerbated income differentiation in less developed countries. In fact, a meta-analysis by Donald Freebairn reveals that 80% of 300 studies on the income effects of the Green Revolution published during 1970-89, found that income inequality increased, both interfarm and interregional (Freebairn 1995). In addition, local food security in many places deteriorated while the national cereal production increased. This can be explained by a shift in production which largely changed from subsistence production to market based production, and from a variety of crops to mainly cereals. Land that previously fed peasants with pulses, was now used for cereal production intended for export (Spitz 1987).

In spite of these concerns regarding the social differentiation of the benefits of the Green Revolution, especially the bias of the Green Revolution for medium- to large-scale farmers remains heavily debated. Vernon Ruttan – for example – argues that the Green Revolution technologies did *not* change income differentiation. Instead, he claims that the situation *before* the introduction of Green Revolution technologies is strongly correlated to the distribution of its welfare effects. Whenever Green Revolution technology was introduced into economies with relatively equitable income distribution it reinforced that equity; when it was introduced into countries with inequitable income distribution in rural areas it reinforced that inequity (Ruttan 2004, p. 14-15). That is not to say that the technology itself had an entirely neutral function in this process. However, the effects on income differentiation were as much related to the existing socio-economic situation, as to the technology itself. Moreover, he argues that in contrast with the mechanisation of agriculture which was biased towards the replacement of labour, the use of improved crop varieties had a predominant land-saving effect, rather than a labour-replacing effect. He concludes that the resulting intensification of agriculture is most likely to have increased both production and demand for labour, leading to an overall positive – instead of a negative – effect on the quality of life in rural villages. In a similar vein, Alston *et al.* argue that some of the negative effects of the Green Revolution in terms of income differentiation may not have been caused by the technology package of the Green Revolution, but by deficiencies in social policies in developing countries. They argue that criticism of the Green Revolution may lead to a revision of research priorities, but more importantly to *"the introduction of complementary policies to address the unwelcome side effects of otherwise beneficial technologies"*(Alston *et al.* 2006, p. 346).

Next to concerns over the social differentiation of the benefits of the Green Revolution, concern has been expressed over the geographic differences in impacts on productivity. For example, Bernstein noted that while the Green Revolution may have been responsible for making a country like India self-sufficient in food grains by the late 1970s, it had a very uneven regional impact. He notes that per capita grain production actually *fell* in 11 of the 15 major states of India, between 1960 and 1985, and was correlated strongly with the distribution of irrigation which enabled multiple cropping (Bernstein 1992). Similarly, the Green Revolution may have had strong impacts in parts of Asia and Latin America, it largely left Sub-Saharan Africa behind (Dixon 1990; Evenson and Gollin 2003). Although a large number of modern varieties have been deployed in this region, the uptake has been minimal, in contrast with Asia and Latin America. One of the reasons may have been that the agro-climatic conditions in Sub-Saharan Africa are less favourable for the type of modernized agriculture that was promoted as part of the Green Revolution. Moreover, high yielding varieties that were available for Asia and Latin America performed poorly in Africa; only in the 1980s did new improved varieties for Africa become available (Evenson and Gollin 2003).

A final major criticism of the Green Revolution is that it has caused severe environmental problems. This is part caused by the poisonous effects of excessive use of chemical fertilizers and pesticides (introduced along with the Green Revolution), and in part because of salination of upper soil layers caused by excessive irrigation (Singh 2000). Moreover, various scholars have expressed the concern that the Green Revolution has strongly contributed to genetic erosion and the large scale replacement of traditional varieties by a limited number of modern crop varieties (Cooper *et al.* 1992; Pretty 1995). This is considered to be a tragic loss of agricultural biodiversity, and therefore the loss of a precious resource for future plant breeding. In addition, a narrow genetic base is feared to increase the vulnerability of cropping systems.

However, also these claims on the negative environmental effects of the Green Revolution remain contested. For example, other scholars have pointed out that the Green Revolution may in fact have helped to conserve environmentally sensitive regions by focusing intensive agriculture on the more productive land, and has reduced the pressure to open up more fragile lands for agricultural production in order to meet the growing requirements for food (Conway 1998; Khush 1999). In addition, Melinda Smale challenges the observations that the Green Revolution is responsible for genetic erosion (at least in wheat), by arguing that the there are different 'windows' or perspectives on genetic diversity. These perspectives range from allele frequencies, to patterns among the plant populations grown on farms in a locality, nation or region. These different perspectives make it very difficult to establish the effects of the introduction of new varieties or genetic recombinations on genetic diversity. Moreover, she argues against the thesis that farming systems have become more vulnerable by stating that there is no evidence for an increased vulnerability of wheat to rust diseases since the rise and widespread use of modern varieties (Smale 1997). Stephen Brush rejects the hypotheses of genetic erosion and instability caused by the Green Revolution for similar reasons, in the context of potato farming (Brush 1992).

Lessons from the Green Revolution, or acknowledging the controversy?

All in all, the Green Revolution has been a very important experience in trying to stimulate large scale agricultural modernisation. It demonstrates the crucial role that improved crop varieties can play in the modernisation of agriculture and the economic development of developing countries. However, it also remains a hotly debated topic. Some forty years after its launch, articles are still appearing discussing its effects on productivity, income differentiation, and the environment. And no consensus appears to be in sight, on any of these aspects. This makes it difficult – if not controversial – to draw lessons from this important and influential past experience in planned agricultural modernisation in the developing world. Paradoxically, it is taken as an illustration of both the great failure of agricultural modernisation (Shiva 1991), as well as its success (Conway 1998).

The continuing attention for the Green Revolution and its evaluation (also in this thesis) may in part be explained by the still relevant questions regarding the validity of the followed strategy for large scale agricultural development. An underlying question in the debate on the Green Revolution is whether technological development can be the key factor in solving widespread rural poverty, or whether such problems need to be addressed by (also) reconsidering social relationships and the distribution of wealth in a society. Donald Freebairn captures the core of this debate by writing:

> *"A technological strategy for agricultural and rural development is politically attractive. If seeds, fertilizer, water control, and pesticides can assure a productive agriculture and a prosperous countryside, the struggles and dislocations of altering social relationships, landholding patterns, political power sharing, and other deeply entrenched arrangements can be avoided. If they cannot, however, other approaches are necessary to help alleviate the destabilizing and demoralizing effects of worldwide rural poverty."*
>
> (Freebairn 1995, p. 277)

In other words, what is at stake in the debate on the Green Revolution is the legitimacy of technology as driver for socio-economic change. As Freebairn writes, a largely technical approach to development would be attractive from the perspective of a social elite that wishes to address rural poverty, without having to get involved in difficult social reforms. However, if the Green Revolution as a largely technological project is evaluated as a failure, it would disqualify such an approach.

Later, in Chapter 3 of this thesis, different theories of technology will be discussed that disqualify a purely technical approach to social change for a different set of reasons. By focusing on the interrelationships between evolving technological and social structures, the notion of technology as external driver will be dismissed. These conceptual discussions on the nature of technology also have repercussions for the evaluation of a project like the Green Revolution, in the sense that they require us to perceive the technologies of the Green Revolution and the social and institutional context in which it emerged as a 'socio-technical ensemble', rather than as two separate spheres.[11] For that reason, it makes no sense to evaluate the impact of technology on an external social reality. Instead, it becomes crucial to understand how the technology fitted the context and motivations in which it emerged, and how that may also explain why the technology was successfully applied in some contexts, while it quite clearly failed to deliver in other contexts. This changes the discussion from a evaluation of good and bad effects of the technology on agricultural production, to a discussion on the importance of developing technology in a contextualized way.

Therefore, rather than concluding this section on the Green Revolution with a definite statement on its value for agricultural development, it is important to acknowledge the controversy, and to acknowledge that the 'success' or 'failure' of a project such as the Green Revolution can only be understood in terms of a specific context of development, and against specific expectations about what the project was supposed to achieve. For example, in terms of national food production, there is little in the way of concluding that the Green Revolution has been a great success, at least for many Asian countries. However, critics have shifted the focus to the socio-economic differentiation of benefits, to food security on household level (instead of on a national level), and to the deruralization that agricultural modernisation has contributed to. In other words, any lessons drawn from the Green Revolution are contingent upon the perspective taken on *what agricultural modernisation is for.*

Leaving the historical perspective of the Green Revolution behind, it is important to shift focus to contemporary developments. The productivity increases caused by the Green Revolution appear to be levelling off (Brown 1997; Strauss 2000), but recent developments in genetics and biotechnology are hoped to provide a new potential to revolutionize plant breeding, and to further increase agricultural productivity. The Green Revolution may have passed by, but a brand new Gene Revolution is dawning (Conway 1998; Swaminathan 2004).

Genetic technologies for agriculture

Genetic technologies play an important role in contemporary efforts to contribute to agricultural development. The previous section has already elaborated the strategic importance of seeds in agricultural production. Moreover, the review of the Green Revolution demonstrated the crucial role of high yielding varieties of wheat and rice in reaching the productivity increases that made the Green Revolution so successful and famous. But while conventional plant breeding may have revolutionized the face of agriculture worldwide, it does have its limitations. Plant breeding is time consuming, and the extent to which breeders can introgress new traits

[11] The notion of a socio-technical ensemble is adopted from (Bijker 1995).

into new varieties is limited. It is not hard to see how this leads to a general interest in the technical potential of biotechnology to overcome limitations in conventional plant breeding.

This general interest in plant biotechnology as a potential solution to problems in agricultural production is fuelled by an increasing public concern on population growth, climate change, and environmental degradation. According to the US Census Bureau the global population is expected to reach a number of 9 billion by 2050, before slowly levelling off.[12] This means that agricultural production will have to keep up with a growing number of mouths to feed. Moreover, global food patterns are changing, with a strong increase in meat consumption in developing countries like China and India (Rosegranta *et al.* 1999). In addition, global warming is expected to have important implications for agriculture because of changing weather conditions, and especially desertification in parts of Africa (Lovett *et al.* 2005). Creating new arable land is considered to be highly problematic, especially if it would mean the destruction of rainforests and other natural habitats. Finally, the search for renewable sources of energy is making the production of biofuels increasingly attractive, which potentially means a competition over arable land between food production and energy production (Ford Runge and Senauer 2007).[13] The international debate on these kinds of global problems rather directly feeds into a plea for the development of technological solutions to the problems in agricultural productivity, especially in harsh environmental conditions.

In order to provide such concrete solutions, the development of new crop varieties through plant breeding is considered to be very important. This is no different from the emphasis on the introduction of modern high-yielding varieties during the Green Revolution. However, one important difference is that the face of plant breeding has fundamentally changed since Mary Dell Chilton led a research group that produced the first transgenic plant at Washington University in 1982 (Pesticide Outlook 2002). This discovery opened the door to entirely novel ways to produce plants with desirable characteristics, and hence provided plant breeding with a new revolutionary potential.

Some technical background on biotechnology, genomics and molecular breeding

In order to appreciate the revolutionary potential of transgenic plants, and other modern plant breeding techniques, it may be useful to provide a little bit of technical background. Please see Box 1.2 for definitions of some concepts that will be frequently used throughout the text.

[12] See: http://www.census.gov/ipc/www/idb/worldpopinfo.html (last accessed 17 September 2008).

[13] Biofuels are alternatives for fossil fuels that are based upon bio-ethanol from carbohydrate rich plants, or bio-diesel based upon oil-rich plants. Common crops used for biofuel production are sugar cane, maize, rapeseed/canola and jatropha.

Box 1.2. Defining biotechnologies.

In this thesis a number of terms are frequently used that may lead to confusion: biotechnology, transgenics, genetic technologies, and modern plant breeding. **Biotechnology** is a commonly used term for the use of biological organisms or processes. The United Nations Convention of Biological Diversity proposed the following definition:

> *" 'Biotechnology' means any technological application that uses biological systems, living organisms, or derivatives thereof, to make or modify products or processes for specific use."*
>
> (United Nations 1992)

This is in fact a very wide definition that may include anything from traditional beer brewing to transgenic technologies or cloning. For this thesis, the definition is somewhat narrower, since the term 'biotechnology' will always be used within the context of agricultural production and plant breeding. However, it will be used in a wider sense than merely to indicate genetic modification (with which the term has sometimes been equalled), and which will be explained below.

The term **genetic technologies** is used in a similar way as 'biotechnology' and denotes biotechnological processes or techniques in which the knowledge, recombination or modification of genetic material is central. This may include genetic modification, genomics, marker assisted breeding or genotyping techniques (which will all be explained in this or later chapters).

The term **modern plant breeding** refers to the practice of making crosses and selections (plant breeding), but with the help of molecular techniques. This may include genetic modification, but more emphatically refers to marker assisted breeding in which knowledge about gene functions, and their traceability through plant crosses can be used to make breeding quicker and more powerful.

Genetic modification is a process in which molecular technologies – rather than a process of natural crossing and selection – are used to specifically alter the content of genetic information (DNA) in an organism. If this process involves the introduction of genetic material from another species, this is called 'transgenics', and hence leads to a **transgenic organism**. Common synonyms for 'genetic modification' include: 'genetic manipulation', 'genetic engineering', or 'genetic transformation'.

Since fragments of DNA (known as genes) are responsible for the expression of a certain characteristic in an organism, the transfer of genes from one organism to another means that also – within certain boundaries – characteristics can be transferred. This is essentially the mechanism underlying breeding, in which the crossing of different organisms is supposed to lead to the recombination of their genetic material in their offspring. However, with the advent of transgenic technology, this recombination of genetic material is no longer restricted to natural crossings (with organisms of the same species), but genetic material can be recombined from entirely different species. A well known example is the transformation of crop plants with a gene from a bacterium *Bacillus thuringiensis*. This naturally occurring soil bacterium carries a gene for a protein with pesticidal characteristics. Transformation of the gene to a crop plant leads to a plant that produces its own pesticide (Vaeck *et al.* 1987). In a similar vein, a rice variety has been produced which produces high levels of pro-vitamin A (*beta-carotene*), with genes from daffodil and a soil bacterium. Because of its yellow colour, this rice variety is commonly known as 'Golden Rice' (Ye *et al.* 2000).

Next to introducing new genes, it is also possible to modify existing plant genes, or to increase or decrease the levels of expression of specific genes. To distinguish these activities from transgenics – in which the introduction of foreign DNA into an organism is essential – it has been named 'intra-genics' or 'cis-genics'. The recent development of a cis-genic strawberry illustrates that naturally occurring DNA fragments can be modified and reshuffled in order to increase disease resistance in crops (Schaart 2004).

The development and application of transgenic crops has been highly controversial and has led to heated public debates, especially in Europe. Important concerns include the outcrossing of transgenic crops with wild species, leading to 'genetic' environmental pollution, and the safety of transgenics for consumption. However, studies into public perceptions of risks associated with transgenic technology have generally provided a very complex and varied picture of why consumers have doubts about the use of transgenics (Marris *et al.* 2001; Frewer 2003). Regardless of what the most important concerns of different stakeholders have been, they have made the development and application of transgenic crops for the European market come to a standstill.

Nonetheless, on a global scale, the use of transgenic crops has been rising for the last 12 years, and big developing countries like India, Brazil, Argentina and China have opened their markets for the commercialization of a limited number of transgenic crops (James 2007). While the first transgenic crops that arrived on the market expressed traits that fitted well within large scale industrial farming in developed countries (e.g. herbicide resistance), today increasing attention is going out to developing transgenic crops that may also have a high relevance for developing world agriculture. Examples include crops with enhanced nutritional value (Dawea *et al.* 2002; Toenniessen 2002), resistance against specific pest insects and viruses (Ferreira *et al.* 2002), improved tolerance to acidic or polluted soils (Herrera-Estrella 1999), and drought tolerance (Moffat 2002). At the same time, in many cases the precise benefits of these crops in developing world agriculture still has to be proven in practice, and will depend on the specific farming system in which they are used.

The development of transgenic crops has perhaps drawn most attention in terms of public debate. However, in terms of technical development, there is a lot of other work being done that is relevant for plant breeding. In the development of transgenics, the focus has generally been on one or a very limited number of genes that are to be transferred or modified for the production of a plant with a new characteristic. The rise of genomics in the past two decades, and most significantly in the last ten years, has shifted focus from single genes, to the functioning of the complete package of genetic material in an organism: to the level of genomes.[14] See Box 1.3 for a brief description of the field of genomics research.

Box 1.3. Genomics, or genome sequencing.

Every cell in an organism contains genetic material, captured in the molecular structure of DNA (*deoxyribonucleic acid*) and written in a genetic code of four 'letters' (nucleotides): A, T, C, and G. A sequence of nucleotides determines what kinds of enzymes are produced by a cell, and by consequence what characteristics a living organism expresses. All genetic material together is called the **'genome'**. Genomics is the science of 'reading' and understanding the complete sequence of nucleotides on the DNA of an organism. By deciphering, or **'sequencing'** the full genetic code of an organism, important insights can be gained of how a living organism 'works', how it can be cured if things go wrong, or how it can be 'improved' (in the case of crops).

Thomas Roderick is said to have coined the term 'genomics' in 1986 to describe the scientific discipline of mapping, sequencing and analyzing genomes, a term that was courteously adopted by the editors of the new journal *Genomics* in 1987 (McKusick and Ruddle 1987). It can be argued that around that time, a new scientific discipline started to take shape and the concept of 'genomics' was born. Most genomics work originally started on micro-organisms (with conveniently small genomes), but the discipline has by far drawn most attention through the Human Genome Project, which aimed to map and characterise the complete sequence of human DNA (Lander *et al.* 2001).[15] However, the implications of genomics for agriculture and plant breeding are most significant in Plant Genomics. Rice and grapevine are currently the only important food crops whose full genomes have been sequenced (Goff *et al.* 2002; Yu *et al.* 2002; Jaillon *et al.* 2007), but efforts are being undertaken to also sequence the full genomes

[14] Consider for example the use of DNA microarray technology, which studies the simultaneous expression of thousands of genes under different conditions at the same time, trying to elucidate – for example – genetic responses to changing environmental conditions on the level of the entire genome.

[15] See also http://www.ornl.gov/sci/techresources/Human_Genome/home.shtml (last accessed 17 September 2008).

of: maize, tomato, potato, cassava, sorghum, castor, soybean, wheat, papaya, Chinese cabbage, banana and a number of other model organisms with less direct agronomic importance.[16]

Plant breeding can benefit from knowledge about plant genomes, because breeders may know better which genetic elements will be responsible for what traits. Therefore they know which genetic elements need to be recombined to get a specific plant with desired traits. Such a plant can then be genetically engineered, but alternatives also exist. Molecular markers on DNA can allow a scientist of plant breeder to relatively easy trace 'genetic elements of interest' in the offspring of a crossing. This allows for a much quicker selection process, and hence for a much larger throughput of crossings and selections. Such 'marker assisted selection' can allow breeders to select for a combination of traits that cán occur through natural crossings, but is statistically highly unlikely to happen (see Box 1.4 for an example).

Box 1.4. The potential of marker assisted breeding – a case of aphid resistant lettuce.

The potential of marker assisted selection is best illustrated with an example: the breeding of aphid-resistant lettuce. Aphids are little insects that feed on lettuce and therefore diminish the commercial value of the produce and require pesticide applications. In the Netherlands, several attempts had been made to develop aphid-resistant lettuce, but this turned out to be very difficult. The trait for 'aphid resistance' appeared to be genetically closely linked to a trait for 'compact growth and rapid ageing', which led to lettuce that was resistant, but with agronomically undesired characteristics. A separation of the resistance trait and rapid ageing trait can occur in normal crosses (through meiotic crossing-over), but it does require a very large segregating population. Moreover, the undesirable 'rapid ageing trait' is difficult to trace visually in a population of lettuce plants, since it is inherited recessively. This means that the presence of the undesirable version (allele) of the gene can be masked by the presence of its desirable version. This in turn makes it difficult to breed and select for a plant with two desirable alleles of the gene, which is required for its commercial use.[1] In this case, molecular markers facilitated the separation of such closely linked traits by quickly recognizing the appropriate recombination of genetic material in new offspring (Jansen 1997).

[1] In technical terms, the recessively inherited allele is difficult to recognize in heterozygous situations, and hence it is very difficult to produce homozygous plants without the undesirable trait.

[16] See http://www.ncbi.nlm.nih.gov/genomes/PLANTS/PlantList.html, or http://www.ncbi.nlm.nih.gov/genomes/leuks.cgi?taxgroup=11:|12:Land%20Plants&p3=12:Land%20Plants for a more extensive list of ongoing plant genomics projects. (Both websites last accessed on 17 September 2008).

From Green Revolution to Gene Revolution?

The potential of transgenic technology, genomics and marker-assisted breeding has revolutionized plant breeding, providing breeders and scientists with the skills and tools to breed crop varieties with an increasingly wide range of potentially useful traits. It is argued to provide projects working on agricultural development with new and interesting options to develop crop varieties tailored to the needs of resource poor farmers, and dealing with the specific problems that arise in their production systems (Delmer 2005; Naylor *et al.* 2005).

In terms of revolutionary technical potential in plant breeding, a comparison between the Green Revolution and the Gene Revolution is tempting, and has been made very explicitly by a wide range of scientists (Conway 1998; Swaminathan 2004; Guerinot 2000). This comparison leads to the assumption that if improved crop varieties were able to boost agricultural productivity during the Green Revolution, the introduction of new improved varieties today can have a similar – or even stronger – impact. At the same time, the technical potential of new crop varieties has to materialize in a world which is also determined by social, political and institutional dynamics and restrictions. This has led to an ongoing debate on the institutional and systemic conditions that may allow or prevent that agro-biotechnological innovations reach resource poor farmers (Tripp 2001; Byerlee and Fisher 2002; Chataway 2005; Reece and Haribabu 2007). In that context, some scholars have stressed that – next to some continuities – there are very important differences in the socio-political landscape in which the Green Revolution and Gene Revolution have taken, and are taking place (Buttel *et al.* 1985; Parayil 2003; Brooks 2005; Swart *et al.* 2007). Rather than assuming a similar effect of the Gene Revolution on productivity, these scholars question the benefits of modern plant biotechnologies within this new context.

The Green Revolution is primarily known as a project of agricultural modernisation, aimed to increase productivity, and to increase food security in the Third World. However, several scholars have indicated that entirely different motivations played a crucial role in supporting this process. For example, Govindan Parayil argues that the main catalysts for the Green Revolution were in fact: (1) strategic considerations during the Cold War to stop the spread of communism in developing countries, (2) the national aspirations of Third World governments to attain food self-sufficiency, and (3) the goodwill of scientists and technologists to contribute to the social and humanitarian goal of eradicating hunger (Parayil 2003). Especially the context of Cold War politics has been frequently mentioned as a crucial factor in this process (Perkins 1997; Hall *et al.* 2000).

> *"Although there was no Marshall Plan to modernize Third World economies, the contingencies of the Cold War prompted the West to find a quick technological fix to avert hunger-led insurrection and possible communist takeover of key Third World nations without demanding drastic changes in the social relations*

> *of production and distribution of their agrarian sector – putatively the crucial economic sector in which most people sought their sustenance."*
>
> (Parayil 2003, p. 986)

In addition, Parayil notes that at the time of the Green Revolution, a strong belief in global modernisation prevailed in international development thinking. Central in this type of modernisation theory is the convergence of modernisation trajectories to a European-style modernity. This ideology of modernisation is argued to have informed the Green Revolution and its strong focus on the transfer of technology from industrialized countries to developing countries (*ibid.*). Interestingly, this approach of planned modernisation seems to fit the institutional backbone of the Green Revolution which strongly relied on international research institutes, funding organisations and national governments, while NGOs or other grassroots organisations played a much less prominent role.

An additional characteristic of the Green Revolution is that it was an entirely publicly funded project, funded by international donor agencies and national governments of developing countries. Market relations and private interests only played a secondary role in the diffusion of the technology. This changed markedly for the more recent Gene Revolution, which is to an important extent private sector-led (Buttel *et al.* 1985; Pinstrup-Andersen and Cohen 2000; Seshia and Scoones 2003). In contrast to the background motivations influencing the Green Revolution, Parayil describes how the Gene Revolution is taking place against a background of economic globalization, in which private sector actors – often multinational corporations – play a leading role in the innovation and diffusion of agricultural biotechnology. He writes:

> *"The technological trajectory is shaped by the imperatives of private property institutions, market forces, global finance, and transnational (and in certain cases national) regulatory institutions. The contingencies and imperatives of economic globalization shape the technological trajectory. New plants and crops are being developed not to solve problems of hunger and deprivation, but mostly to increase shareholder values of companies that have invested heavily in R&D efforts in the biotechnology sector."*
>
> (Parayil 2003, p. 982-983)

This marks a crucial difference with the dynamics of the Green Revolution, that were primarily led by national governments and international donor agencies. Rather than geo-political interests, commercial incentives appear to be determining the development and diffusion of modern plant varieties. This also has repercussions for public sector research. Sally Brooks argues that structural underfunding of public sector agriculture research institutes means that agro-biotech development is mainly determined by the private sector, and that public sector research institutes are increasingly reliant on public private partnerships (Brooks 2005). This new playing field, with a leading role for the private sector, is further shaped by the rise of increasingly important and strict intellectual property regimes to protect private interests in biotechnology development (Dutfield 2003). It provides a legal structure for the new socio-political landscape in which modern biotechnology and plant breeding is taking place. Unfortunately, this intellectual property regime is feared to further restrict the 'freedom to operate' of the already underfunded public sector (Falcon and Fowler 2002; Atkinson *et al.* 2003).[17] Box 1.5 provides a summary – adopted from an article by Sally Brooks – of the main continuities and differences between the Green- and Gene Revolution.

Box 1.5. Continuities and changes between the Green Revolution and Gene Revolution (Brooks 2005: 362).

Continuities
- Promotion of 'scientific revolution' in agriculture; a 'technological fix' applied to complex socio-economic realities.
- Promotion of monocultures to intensify production.
- Food shortage presented as a supply problem rather than a distribution problem.
- High barriers to entry tend to squeeze out smallholders and increase inequality.
- Legitimized by neo-Malthusian discourses.

Changes
- High levels of uncertainty and risk surrounding transgenic technologies, new issues such as bio-safety.
- Ownership and control: from public sector to private sector.
- International context: from Cold War and national food self-sufficiency to neo-liberal globalisation and competitive exports.
- A wider range of actors influencing and contesting policy.

The picture is clear; although there are some clear continuities between the approaches of the Green Revolution and Gene Revolution, crucial differences have been described in terms of the driving forces behind these 'revolutions', and the rules of the game. The result is that the Gene Revolution is taking place in an entirely different playing field than the Green Revolution. Modernisation as overall ideology of development is increasingly being reshaped, reinvented and legitimized by a globalizing economy and the interests of a powerful private agro-biotech industry.

This discussion on the motivations and drivers behind agricultural modernisation during the Green Revolution and in more recent times, has important consequences for the analysis of

[17] The tension between an increasingly strict intellectual property regime and ways to increase freedom to operate is further elaborated in Chapter 3.

reflexivity in technology development projects, as introduced at the beginning of this chapter. It indicates that political, ideological and commercial motivations may have an important influence on the nature of development projects and processes. This leads to the conclusion that contemporary projects on agricultural development and their reflexivity cannot be evaluated in a narrow sense. They too need to be evaluated with respect to the contemporary socio-economic situation, its discourses, and its different interests.

Making genetic technologies 'appropriate' for agricultural development

While the previous section highlighted the historical and socio-political motivations that may influence processes of agricultural development, these issues rarely penetrate the mainstream discourse on agro-biotechnology development for the poor. Instead, the political, commercial or ideological backgrounds to development generally remain opaque, and the purely humanitarian motive of 'helping the poor' is put to the fore, both by public and private sector actors. But regardless of its precise motivations, every project needs to make its technology work in a new and difficult environment. This has led to the common-sensical – but rather depoliticized – notion of 'appropriate technology'. This notion of appropriate technology emerged from the general acknowledgement that not any transfer of technology from an industrialized country to a developing country is successful. Important contextual factors play a role in determining how new technologies interact with a local production systems. Appropriate technology is supposedly a technology that fits well within the local circumstances in which it has to perform. The question remains how to determine what makes technology 'appropriate'.

This question of how to define 'appropriate technology' is not new. It has been posed for several decades in international development debates, and goes back to the early 1970s (Shumacher 1973). Despite of the length of the debate on 'appropriate technology' a consensus on its meaning seems to be lacking, and very different conceptualizations circulate in mainstream discourse and in various articles. Definitions commonly vary from appropriateness in terms of adaptation to local climatic conditions, socio-economic conditions, cultural preferences, and market opportunities. But it is also defined as:

> *"Applied science that is suitable for the level of economic development of a particular group of people. Appropriate technology is decentralized, can be understood and operated by its users (i.e., does not require outside operators), uses fuel and other resources that are either local or easily obtained, and involves machinery that can be maintained and repaired by its users. Often, but not necessarily, it is labor-intensive and involves simple machinery."*
>
> (Art 1993; The Dictionary of Ecology and Environmental Science).

A similar, somewhat shorter quote that is circulating on the internet, and is attributed to British architect John FC Turner, reads:

> *"Appropriate technology is technology that ordinary people can use for their own benefit and the benefit of their community, that does not make them dependent on systems over which they have no control"* [18]

What these definitions have in common is the focus on 'appropriateness' not only in the sense of technical functioning, but in the sense of social relations of production and maintenance, and a relative independence from external inputs. This provides a starting point for a somewhat richer notion of appropriateness. In such a notion, the question of appropriateness moves beyond technical considerations alone, but explicitly questions for *whom*, and for *what kind of development* a given technology may be appropriate.

The argument to contribute to appropriate technology development has commonly led to a plea for bottom-up agricultural technology development. The basic line of thought is that appropriate technology development should start with locally defined needs and priorities, and take on board locally relevant knowledge and preferences. This is practically organised by actively involving farmers and other local stakeholders in priority setting exercises, and the evaluation of technical solutions that are being developed. Such notions of participatory biotechnology development have for example been elaborated by Joske Bunders and Jacqueline Broerse (Bunders 1988; Bunders and Broerse 1991; Broerse 1998; Broerse and Bunders 2000) and link up with a range of other methodologies for participatory agricultural innovation broadly captured under terms such as 'Rapid Rural Appraisal' and 'Participatory Rural Appraisal'.[19]

An important implication of this strategy is that technological development becomes demand driven (instead of science driven), and hence supposedly appropriate to the context of application. Moreover, this use of participatory methodologies explicitly answers the question of *who* the beneficiaries should be of technology development: appropriateness is defined with respect to the participants of the priority setting exercise. However, there are other questions that run a risk of remaining implicit and unquestioned in this kind of methodology. These questions relate to the kind of agricultural modernity that technological development is supposed to lead to, and especially to the social relations and responsibility in processes of agricultural innovation. This risk is especially evident if participatory methodologies would only interact with the local and micro-level specificities of a given project. Such an approach may gain validity on a local level, but at the same time runs the risk of obscuring that agricultural development can have important long term consequences for the way in which

[18] See e.g. http://en.wikipedia.org/wiki/Appropriate_technology (last accessed 17 September 2008).

[19] See Chambers (1994) for a introduction and overview of experiences with RRA and PRA. See "Farmer First" (Chambers *et al.* 1989) for the seminal classic on involving farmers in agricultural innovation. Niels Röling has been another important proponent of participatory learning in agricultural development. See e.g. Röling and Wagemakers (1998).

farmers are linked up with markets and production chains, how seed systems are organised, and how agricultural production itself is being organised. Hence Mohan's argument that *"Most participatory approaches tend to study down to the local level, but more transformative approaches would also study the global economy and transnational organisations such as the major development agencies and be prepared to criticize bad practice."* (Mohan 2001, p. 164).[20]

In spite of such concerns, stakeholder involvement in technological development is an important way of making technologies more 'appropriate'. It is a particularly good way of learning about local priorities, and about defining the direct beneficiaries of development. However, the concerns raised in the previous paragraph imply that it is insufficient to delegate all questions regarding the future development of agricultural production system to participatory exercises, or to legitimize processes of agricultural development merely by reference to locally defined priorities. Instead, a critical reflection upon the mode of agricultural modernisation, and the position of farmers in that process is required. Such a critical reflection may clearly be a part of well-balanced participatory methodologies, but is not sufficiently guaranteed by stakeholder involvement alone.

In summary, the notion of appropriateness is ambiguous, and depends not only on technical parameters, but also on the questions *'appropriate for whom'* and *'appropriate for what kind of agricultural modernity'*. This complexity of appropriateness is problematic, but at the same time it may provide an interesting and useful empirical entry point for analyzing and comparing different approaches to agricultural development. Every development project must have found a way of dealing with this issue, and must have made implicit and explicit choices on how to make sure that its technological outputs are 'appropriate' solutions for the problems it is addressing. This means that the notion of appropriateness – in all its vagueness – provides an excellent entry point to researching the reflexivity of development projects, what the crucial criteria are considered to be for making development work, and what the relationships are with wider trends of agricultural modernisation. And that is precisely what this research aims to do.

Concluding remarks

This introductory chapter started by discussing the ambiguity in international development efforts, and by introducing the notion of reflexive development which highlights the process in which development projects or institutions reflect upon their own work and respond to tensions and criticisms generated by that work. Rather than dividing the development sector in various rigid approaches with their respective ideological underpinnings, this notion highlights flexibility and a learning dynamic which arguably more accurately describes how

[20] See also Cooke and Kothari (2001) for a wider selection of criticisms and limitations of participatory methodologies.

development efforts have changed over the years and have – for example – been influenced by the work of civil society organisations.

The consecutive discussion of agricultural modernisation efforts in the past, and the comparison between the 'Green Revolution' and the 'Gene Revolution' illustrated that the context of agricultural development is characterised by rather strong controversies over what effects of agricultural modernisation in the past have been, and what the role of technology transfer can or should be. Moreover, it illustrated that the nature of such development efforts may be strongly connected with a wider context of geopolitical considerations, or with an increasingly important role of the private sector in the area of plant breeding and biotechnology development. This reaffirms the importance of reflexivity in agricultural development efforts, not only in relation to its direct costs and benefits, but also with respect to the wider historical trend that such processes of agricultural development are part of.

These questions regarding the nature and effects of agricultural development efforts come together in the notion of 'appropriateness' of development projects and the technologies they develop or use. As discussed, this notion of appropriateness is highly ambiguous and can relate to anything ranging from very instrumental considerations to more profound questions regarding the social structures that new technologies require or create. This means that the concept is fundamentally problematic, but at the same time may serve as a conceptual entry point in order to investigate different projects of agro-technology development and the way in which appropriateness is defined and operationalized in practice in these projects.

With this introduction, a starting point has been provided for the research presented in this thesis, including a number of key concepts that will provide a framework to question and analyse various projects of pro-poor agro-technological development. These concepts – reflexivity, appropriateness – will be further discussed and defined in Chapter 3 against a context of agricultural modernisation and industrialisation. That chapter will reflect upon trends in agricultural modernisation and industrialisation that have prescribed a very specific path of agricultural development in which seed breeding and other types of innovation have become externalized and have been transformed into industrial inputs for the farming system. The question will be raised to what extent such a model of agricultural development is still legitimate, where it runs up against its limits, and especially whether alternatives can be envisioned. This discussion will provide a more complete conceptual framework and background for studying concrete examples of 'pro-poor' agro-technological development in Chapters 4 to 6 of this thesis. Before that, Chapter 2 will first elaborate on the research design, methodology and the main research questions for this study. But first, this chapter will round up by discussing the scientific and social relevance of this study, and by providing an overview of the structure of the thesis with a brief summary of the topics discussed in the various chapters.

Scientific and social relevance

The thesis engages with debates on agricultural development, but especially with debates on the role of technological innovation as part of processes of agricultural modernisation. It aims to widen the debate on 'appropriate biotechnology development for resource poor farmers' by reflecting critically on the way in which technical innovation relates to changing and existing social relations in agricultural production. This means that the analysis in this thesis has an important scientific relevance. In bringing together development studies, science and technologies studies, and critical theory, it creates a critical – but constructive – perspective on technological development and modernity in the context of international agricultural development. While science and technology studies have stressed ambiguity and contingency in the construction of technology, critical theory and development studies have generally taken a wider historical perspective and stressed persistent historical trends in the relationship between technology and international social order (Edwards 2003). The contribution of this thesis lies in highlighting the relationship between both types of analyses, and in focusing on the relationship between technical development, and social relations in agricultural development.

In practice, this also means that the study explicitly seeks a confrontation between conceptual frameworks, influenced by science and technology studies and critical theory, and the empirical reality that emerges from case study analysis. Conceptual notions such as 'the politics of technology' and 'externalization as part of agricultural industrialisation' which will be introduced and elaborated in Chapter 3, will be used for concrete empirical analysis of contemporary projects of technology development. This confrontation between conceptual and empirical levels not only allows for a novel and potentially insightful analysis of these projects, it also implies a crucial test of the validity and usefulness of these concepts in a practical setting. Therefore, the final chapter of this thesis will reflect upon the usefulness of the conceptual framework that has been adapted for the case study analysis, and the need for adaptations or refinements of these conceptual notions. This will in particular result in a discussion of the relationship between technological design and wider social structures in a practical setting, as well as in a discussion of the contemporary understanding of the notion of 'appropriateness' in the context of international agricultural development.

Finally, in addition to its scientific relevance, the analysis in this thesis is motivated by a clear social engagement and as such aims to have a strong societal relevance. The potential impact of international agricultural development is significant, as demonstrated by the effects of the Green Revolution. However, as the same Green Revolution demonstrates, there is an urgent need to move beyond a discussion on the 'means' towards agricultural modernisation, and to reflect upon its 'ends', in terms of what agricultural production systems may look like in the future, and how benefits of agricultural modernisation are differentiated. A more sophisticated understanding of what appropriate agricultural technologies may be, and how they can be defined in a concrete setting, can potentially increase their impact. Moreover, considering ongoing discussions on the institutional structure and the mandate of – for example – the CGIAR system (CGIAR Secretariat 2008), a richer and contemporary understanding of what appropriate technology development could mean is expected to contribute policy-relevant insights. Such insights will be further discussed in the final chapter of this thesis, and will in particular relate to the position of (partly) publicly funded development projects vis-à-vis a wider range of actors in the agricultural innovation system.

Structure of the thesis

This first chapter has been committed to introducing the area of interest for this thesis, the dilemmas in agricultural development for resource poor farmers, and the question of how to operationalize the notion of 'appropriate technology' for agricultural development. This has provided us with a starting position for formulating concrete research questions, and for the description of three cases in the upcoming chapters.

Chapter 2 describes the research design of this study, and in doing so it introduces the main aim, research questions and methodology for data collection of this research. As will be explained, the research is strongly exploratory in nature and therefore – next to a literature study – it primarily relies on qualitative research methods and case studies for its data collection and analysis.

Chapter 3 will provide a more extensive conceptual background, deepening the discussion started in the first introductory chapter. Agricultural development is discussed in terms of modernisation and industrialisation processes which are argued to both contribute to a relatively homogeneous approach to agricultural development, and to the externalization of many aspects of farming practice like breeding and seed management. Against this background an interest is expressed in the possibility to use genetic breeding technologies, without necessarily externalizing agricultural innovation to specialized breeding institutes or companies. This focus is legitimated by a perceived tension between a standardized package of agricultural advice and technologies, and farming systems in difficult environments which are characterised by a high degree of variability and localized adaptation. It is suggested that farmers in such areas are likely to require a more open-ended approach to agro-technological development in which they are empowered in their own on-farm experimentation with new crop varieties. The question is whether elements of such an approach to agro-technological innovation can be witnessed in the case studies, and what that means in practice for technical design.

Chapter 4, 5 and 6 present three case studies of projects in which plant breeding and genetic technologies are used to develop new crop varieties with interesting traits for resource poor farmers in developing countries.

Chapter 4 presents the case of the Collaboration on Insect Management for Brassicas in Asia and Africa (CIMBAA); a public private consortium in India which aims to develop a cabbage variety which is resistant against the diamondback moth. This insect is currently causing big losses in cabbage cultivation in India, and the CIMBAA consortium hopes to address this problem by engineering Bt insect resistance into a cabbage variety. The case study touches upon several dimensions and aspects of making genetic technology appropriate for resource poor farmers, including the technical design of the gene construct that is used, the structuring role of intellectual property in the consortium, and the scope of stakeholder involvement in this project. The case is taken as a main illustration of how extensive efforts to reach resource poor farmers remain within the limits of an already existing industrial production system, in which the role of an external seed supplier is legitimized and consolidated. The innovation process in this case is characterised by its treatment of farmers as recipients of technology, and by their indirect representation in the project, rather than by their direct involvement.

Chapter 5 presents a set of initiatives of the International Potato Centre in Peru (CIP). The Peruvian Andes are the centre of origin of potato, and traditional potato production is characterised by the use of a wide diversity of landraces. The use of modern improved potato varieties may boost productivity for farmers, but is feared to lead to the replacement of these native potato varieties, which are an important resource of genetic diversity for future plant breeding, as well as an important private resource for Andean potato farmers. For this reason, CIP is experimenting with participatory breeding programmes, the repatriation of native potato varieties, and the marketing of traditional potato varieties. These initiatives are argued to challenge the common bias in agricultural modernisation towards a narrowing genetic base, and the specialization on a very limited number of crop varieties. In addition, the centre is experimenting with virus resistance kits, which may significantly slow down the degradation of potatoes because of virus infestation. The combination of improved virus resistance of potatoes, diagnostic techniques and improved virus management practices may allow farmers to sustainably produce their own seed potatoes, providing them with a reasonable alternative to commercially available seed potatoes. The case study highlights how the technological interventions by CIP are capable of challenging (at least on a conceptual level) ongoing trends towards an industrialisation of potato production, and empowering farmers in their own on-farm seed potato production. Finally, the case represents an example of how farmers can be involved in agricultural development as co-innovators with specific valuable and complementary knowledge and expertise.

Chapter 6 presents the work of the CGIAR Generation Challenge Programme (GCP), which is committed to the use of upstream comparative genomics research for the development of drought resistance traits in crops of interest to resource poor farmers. A main difference with the other two case studies is that the outputs of this research project are so upstream that an evaluation of their social meaning in a concrete production system is not meaningful. By consequence, the case study focuses on how the GCP aims to link upstream genomics research to concrete development objectives in developing world agriculture. The chapter evaluates the priority setting exercise conducted by GCP and the operationalization of an innovation chain perspective, aimed at making sure that the outputs of GCP research are actually taken up by downstream research partners. The chapter discusses some of the potential difficulties in this approach and explores the potential for complementary innovation systems in order to meaningfully link upstream science-led genomics research and downstream bottom-up breeding programmes. This exploration address the various partnerships of the generation challenge programme, as well as the Genotyping Support Service (GSS) as a specific technical interface between upstream genomics research and downstream variety development. The GSS is a very accessible service which allows the outsourcing of molecular analyses for a variety of projects. This initiative is taken as a potentially very interesting approach to agro-technological development that shifts focus from the development of a technical solution, to the provision of a technical service. As such, the case is a clear example of treating local research partners and farmers as co-innovators in agricultural development.

The burden of Chapter 7 is to bring together the analyses of the three case studies and to evaluate how the different projects have practically operationalized the objective to develop 'appropriate technology' for the agricultural development of resource poor farmers. This leads to an extensive discussion on the different dimensions in which appropriateness of technological innovation is interpreted and reconsidered, and to the formulation of a contemporary understanding of what appropriateness means in practice. The multi-dimensional understanding of appropriateness that emerges from this analysis is taken as an argument for 'reflexive biotechnology development' as an approach to agro-technological innovation. The chapter further reflects upon the extent to which the material design of the various genetic technologies in the case studies is related to specific structures of production or innovation systems, and the extent to which the use of genetic technologies in plant breeding necessarily leads to an externalization of the innovation process. Some practical implications of the study for contemporary innovation policy are discussed and new questions for future research are formulated.

Chapter 2
Research design

> *"It is ... the posture of the constructivist paradigm that there is no single 'real' reality, but only multiple realities constructed by human beings."*
> (Guba and Lincoln 1989, p. 64)

> *"Reality is what we choose not to question at the moment"*
> (John Dewey, quoted in Becker (1993, p. 219))

Motivation and objectives of the study

As has become clear from the introductory chapter, the research presented in this thesis engages with debates on reflexive development and appropriate technology, but aims to deepen these debates by reflecting upon the relationship between contemporary technology development projects, and a wider context of agricultural modernisation and industrialisation. As has been discussed in Chapter 1, there are various dimensions in which to consider 'appropriateness'. Moreover, the discussion on how both the Green Revolution and the Gene Revolution are influenced by wider socio-political trends and interests, has made clear that appropriateness can only be understood in reference to a wider socio-political and historical context. All in all, this illustrates that an apparently simple question of how to make technological development responsive to local needs and circumstances, is in fact a very complex and layered question. It requires a reconsideration of what can make technology 'appropriate', for what kind of development and for whom. For this reason, the research presented in this thesis will have to transcend a merely technical approach to 'appropriateness', and instead ask whether also a reconstruction of social relations of production and innovation are possible within current projects of pro-poor agro-biotechnology development. It will reflect upon the way in which genetic technologies are co-evolving with a specific perspective on agricultural modernity, and the various roles that scientists, breeders and farmers play in that future.

In summary, this research is motivated by the concern that agro-technological development may be focused on taking local priorities and conditions as starting point, but that it hardly reflects on how social relations in agricultural innovation and production are changing as part of the development process. Concretely, and as further discussed in Chapter 3, this primarily concerns the legitimacy of the ongoing externalization of innovation as part of agricultural development. As such, it relates to the question whether farmers and other stakeholders are primarily conceptualized as recipients of technology, or as co-innovators in the innovation process. The aim of this thesis then is to bring this level of questioning technologies for agricultural development to the surface, and to discuss how in different projects of agro-technological development the notion of 'appropriate development' is operationalized in order to contribute meaningfully to agricultural development for resource poor farmers.

Research questions

Concretely, the above translates into the following main research questions, which will be leading in the analysis of three case studies that will be elaborated in this thesis. The first is the most general, and allows for a concrete entry point for the analysis of case studies:

> *How do contemporary projects of pro-poor agricultural biotechnology development operationalize their pro-poor focus, and what criteria are – implicitly and explicitly – taken on board in that consideration?*

This general research question is made more specific by the following research question, which emphatically focuses on the social relations of production and innovation. This second research question clearly moves the focus beyond an instrumental analysis of pro-poor technologies, and their 'appropriateness' for resource poor farmers:

> *How are farmers conceptualized as end-users in the operationalization of appropriateness, and what does that mean for their involvement in the innovation process and for their position in the future production process?*

These main research questions provide a general entry point for the case study analysis in the upcoming chapters. More specifically, they invoke the following five 'study questions' that allow a direct investigation of the three cases presented in this thesis.

1. What is the background of the project? Why was it set up and what are its main objectives?
2. What kind of technology is being developed and applied to reach the project's goals? How is that technology adapted and modified to fit the specific project goals and the context in which it has to operate? Which criteria are followed in order to make the technology appropriate?
3. In what other ways is the project making an effort to reach its objectives? What dilemmas does it encounter and how does it solve those dilemmas?
4. What efforts are being made to involve farmers and other end users or stakeholders in the process of technology development? What image of these stakeholders is constructed in order to grant them a more or less important role in the project?
5. What kind of agricultural production system is implicitly and explicitly supported by the project? What are the different social roles and responsibilities in that production system for farmers and technology developers?

Together, these research questions provide a tool to critically investigate contemporary projects of pro-poor biotechnology development, linking the concrete interpretation of what constitutes 'appropriate technology' to a wider trend of how social relations in agricultural development are changing.

Approach – explorative qualitative research

The study proposes an investigation of different ways of operationalizing the notion of 'appropriateness' of technologies in practice. Moreover, this analysis is supposed to shed light on both explicit and implicit criteria, assumptions and considerations that are taken on board in the development of 'appropriate' genetic technologies. Rather than a superficial inventory of the formal considerations for technological design, this requires a detailed view and analysis of concrete projects, their history, rationale, considerations and technical outputs. Moreover, the research questions that are being asked inquire into the nature of the process of agro-technological innovation, and hence require a qualitative analysis – rather than a quantitative analysis – of different contemporary projects of pro-poor agro-biotechnology development. In order to achieve such an analysis, the study of a select number of cases of pro-poor biotechnology development is considered to be a suitable research methodology (Yin 2003). The careful analysis and deconstruction of the criteria they apply in technology development will provide insight in the various ways in which 'appropriate pro-poor technology' is operationalized.

While the case study is adopted as the primary research methodology, these case studies do require an embedding in a wider context, as well as a theoretical framework in order to allow for a sharp and insightful analysis. For this reason, the research has started with an extensive literature review in order to obtain insight in the contemporary and historical context of agricultural development and plant breeding. This literature study allowed for the positioning of the case studies against a wider background and for a perspective on the long term trends in agricultural innovation. The findings of this literature review have constituted the basis of the introductory chapter of this thesis, and are further elaborated and discussed in Chapter 3. In addition, the literature study has contributed to the development and refinement of a conceptual framework. This conceptual framework is also further explicated in Chapter 3, and is crucial for the specific case study analysis that has been undertaken. Especially the strong relationship between technological design and social structures that is discussed and questioned in the conceptual framework has informed the questions for the case studies, as well as their analysis.

Case study selection

Three case studies – elaborated in Chapters 4 to 6 – constitute the empirical heart of the research presented in this thesis. They include (1) a study of a public private consortium in India that develops transgenic insect resistant cabbages for Indian vegetable farmers, (2) a study of various initiatives of the International Potato Centre in Peru, and (3) a study of the Generation Challenge Programme that invests in genomics research into drought tolerance. Figure 2.1 provides a quick overview of these three cases. Much more extensive case descriptions are provided in Chapters 4 to 6 as introductions to the individual case studies.

The case studies were deliberately chosen to represent different institutional settings. In selecting such different settings, two criteria for categorizing different projects immediately come to the fore: their nature of funding (public or private), and their relative position in the

<table>
<tr><th></th><th>Public</th><th>Public - Private</th></tr>
<tr><td>Upstream</td><td>Theme:
Comparative genomics for drought tolerant crops.
(CGIAR's Generation Challenge Programme)

Institutional context:
Publicly funded international research programme

Main challenges:
How to design an impact strategy for upstream genomics research, while being open for bottom-up priority setting?</td><td rowspan="2">Integration of Upstream-downstream phases within one project

Theme:
Transgenic cabbages for resource poor farmers (CIMBAA Consortium, India)

Institutional context:
Consortium with one private seed company and various public sector institutes

Main challenges:
How to balance private seed company's interest with the wish
seed/technology for resource poor farmers?
How to legitimize the use of transgenic technology for resource poor farmers?</td></tr>
<tr><td>Downstream</td><td>Theme:
Potato breeding in the Andes (International Potato Centre, Peru)

Institutional context:
Publicly funded international research institute

Main challenges:
How to provide useful new varieties without jeopardizing crop genetic diversity in the centre of origin of potato?
How to link formal wi
systems?</td></tr>
</table>

Figure 2.1. Overview of the case studies. The analysis of these three case studies covers different types of projects in terms of their funding, as well as in terms of their phase in the upstream-downstream continuum of innovation. The grey numbers printed in the background of the table above indicate the order in which the case studies are presented and discussed in this thesis.

process of innovation (upstream or downstream). The three case studies chosen represent public and private perspectives, as well as upstream and downstream phases in development process. These criteria are further elaborated in the following two sections.

Source of funding as criterion for selection

The public or private source of funding was mentioned as a first criterion to select different case studies. This public private dichotomy is not an absolute criterion, since sources of funding for many pro-poor biotechnology projects are diverse and come from both public and private sources. In fact, rather than the sources of funding being relevant themselves, they serve as an indicator for the interests underlying the project. The crucial difference here is between projects that are primarily public in nature and have a merely humanitarian focus, and projects with a substantial amount of private funding and an element of (future) commercial development.

In the field of international agricultural development, the CGIAR still is one of the major public sector protagonists, doing breeding work and agricultural research in 15 centres worldwide, and with a clear pro-poor focus. The CGIAR receives funding from an international donor community that consists of both governments and private funding organisations. While this donor community lays down some requirements in terms of demonstrated impact of the funded research, the CGIAR does not have any commercial objectives in its technological development work. The only stake the different institutes have is 'staying in business', by making sure that their contribution to international agricultural development has an impact, and is considered to be legitimate by its donor community. In terms of case study selection, projects from the CGIAR are considered to be representative of a public-sector approach to agricultural development in which the technology developer has no significant future stake in the production process. Concretely, the CGIAR's Generation Challenge Programme and the International Potato Centre are selected as case studies representing the public sector perspective.

In order to contrast public sector case studies, it would be interesting to study an entirely private sector project as well. However, projects that are entirely owned and funded by private sector companies, and aim at resource poor farmers in developing countries are rare. Although developing countries are increasingly recognized as potential growth markets, seed- and biotechnology companies primarily focus on the commercial segment of such markets, and the relatively rich farmers. The development of seeds or genetic technologies for resource poor farmers is generally not considered to be remunerative for a private sector company alone.

Having said that, there is an increasing attention for public private partnerships in the area of pro-poor biotechnology development. In such partnerships, the presence of public money and support may make investing in seed and technology for resource poor farmers interesting for seed companies. In addition to a public interest in supporting such projects, participating companies may have a commercial interest in securing future seed sales, or in providing a proof-of-concept for their seed technologies in difficult environments. For public sector research institutes and universities, such public private partnerships can be valuable because of the expertise of a private sector company, or because the partnership grants the project access to proprietary technologies of the company that would otherwise be hard to use. In terms of case study selection, this means that purely private projects are not taken on board in this research, but that a public private partnership has been selected as a counter perspective to the public CGIAR cases: the CIMBAA consortium in India.

Upstream-downstream as criterion for selection

Next to the nature of funding, the relative position of projects in the process of innovation is another important criterion to categorize different types of projects. An important distinction can be made between 'downstream' projects that are engaged in the very concrete development of new crop varieties, and in 'upstream' projects of genomics or biotechnology development that do not have an immediate relevance for farmers, but can provide important inputs for downstream breeding programmes. Like in the public private dichotomy, this distinction between upstream and downstream is by no means absolute. Some projects aimed at the development of a concrete new crop variety may also involve upstream genetic research, which directly leads into a new crop variety. Especially in the case of the development of transgenic crops, an implosion of the upstream-downstream continuum can be observed, and a single project may require the integration of both types of research and development. Still, in general terms, the nature of research, the institutional context, and interactions with other stakeholders can be markedly different for upstream and downstream projects. While upstream projects can generally afford to be more isolated and science-led, projects of downstream variety development are generally characterised by a much more intense interaction with farmers, regulatory bodies, and civil society in general.

Final considerations for case study selection

These have been the most important selection criteria for the three case studies presented in this thesis. In addition, the selected case studies were considered to be especially interesting because of specific challenges that the projects were dealing with. Challenges for a project may for example arise out of conflicting requirements of funding bodies, out of public controversy over new technologies, or out of different interests of organisations within a consortium. Next to making these case studies simply more interesting, such challenges within studied projects provide an analytical advantage. A clear vision on a common objective can provide a way out of dilemmas or tensions within a project. For this reason, the way projects or organisations deal with challenges in their work provides an excellent entry point to study both their explicit as well as their implicit goals and objectives, and the background reasoning which makes their work legitimate and worthwhile. It uncovers what arguments and considerations are most important for a project, and what kind of considerations are treated as 'irrational' or 'irrelevant'.

Finally, some more pragmatic considerations have played a role in the case study selection for this study. Key requirements for the case studies included an explicit focus on agricultural development for resource poor farmers in developing countries, and the use of genetics for the development of new crop varieties or any other agricultural input for those farmers. Based upon these criteria, and apart from the cases selected for this study, a wider range of CGIAR projects might have been selected that would have been representative of public sector research. Similarly, a wider range of public private partnerships could have been chosen from, like for example the ongoing ABSP II projects.[21] A decisive reason for the selection of the CIMBAA project (rather than another public private partnership) as a case study in the context of public private partnerships was the accessibility to the project and its stakeholders. In comparison to other public private partnerships contacted, it was relatively open to investigation by the researcher which provided an essential precondition for a rich case study description and analysis.

In the context of publicly funded research projects, the Generation Challenge Programme is rather unique in its research focus on upstream genomics research with a clear downstream development objective. In that sense, although other public sector research programmes exist, this characteristic made the programme particularly attractive as case study. Finally, the CIP case study was selected from a wider range of potential public sector case studies because of the expectation that the work with small scale farmers in the Andes, in the centre of origin of potato and with many issues related to the potential loss of biodiversity would create interesting challenges for the research institute, perhaps inviting novel and creative approaches to making sure its work is 'appropriate' for its intended beneficiaries. At the same time, other projects or institutes could have been selected for similar reasons. The CIP case study is regarded to be generally representative for a wider range of public sector plant breeding initiatives, although the case study in Chapter 5 will extensively focus on the specific and unique factors that shape the work of this institute in Peru.

Data collection – a technographic approach

As has become clear from the methodological approach elaborated in the previous sections, the concrete empirical work for this thesis is to analyse and deconstruct the stories of how specific technological projects have been setup and how they are responding to their own

[21] ABSP stands for 'Agricultural Biotechnology Support Project', and is a Cornell University hosted project with various sub-projects that involve both local public sector partners as well as private sector partners. The project is essentially aimed at the transfer of biotechnology from the private sector to public sector partners. See http://www.absp2.cornell.edu/ (last accessed 15 December 2008) for more information about this program.

project goals and the context in which they have to operate. This invites a research approach that is akin to ethnography, but with a specific focus on technological systems and the social actors involved in those systems. Concretely, a methodological approach that is useful in that respect is 'technography'. Steve Woolgar (1996, p. 89) describes technography as *"the social-scientific study of technical settings"*. He argues that technography adopts certain features of the 'ethnographic' method in that it observes and describes social behaviour in a particular setting. Importantly, the ethnographic method emphasises the need to resist taking for granted the various categories and characterizations used by the people being observed, such as the distinctions made between 'producer', 'consumer' and 'user'. He argues that in line with the ethnographic method, *"a main focus of technography is to determine how these distinctions are created and sustained, as well as determining what effect they have on design and development"* (*ibid.*). Concrete and recent examples of technographic studies are provided by Zannou (2006) and Kassiwike (2008), who for example described this approach as *"an attempt to map the actors, processes and client groups in such a way that the analyst can see beyond the technology itself, to the problems technological applications are supposed to solve and to understand what parties and interests are being mobilised in arriving at solutions"* (Kassawike 2008, p. 22-23).[22]

These general objectives of the technographic approach invite data collection from a very wide range of resources, which may include interviews, informal conversations, and the study of a wide range of policy documents, websites and other media. All these resources may contribute important information to the researchers understanding of the technological project, and the way in which various 'parties and interests are being mobilised' in order to arrive at solutions, or more in general to provide the project with legitimacy. However, one of the important assumptions underlying the data collection for this research is that the setup and development of a project of technology development is not unambiguous. As Guba and Lincoln (1989, p. 64) put it: *"It is ... the posture of the constructivist paradigm that there is no single 'real' reality, but only multiple realities constructed by human beings."* By implication, it is anticipated in this research that instead of a single 'formal history' of a project, in practice a range of parallel and possibly competing versions exist of how a project is setup, what its rationale for existence is, how it interacts in practice with various stakeholders, and to what extent it is delivering useful outputs that allow the project to reach its goals. In fact, as will become clearer from the case studies presented in Chapters 4 to 6, the projects studied and the technology developed cannot be seen in isolation from a wider discourse on agricultural development, and specific debates on the use of transgenic crops, the loss of biodiversity and the industrialisation of agriculture. This implies that the representation of what a project is, and how 'well' it is doing is almost by definition contested terrain, and should be considered against the wider context in which a project is working, as well as against the individual and institutional interests that may be at stake.

[22] Both Zanou and Kissawike refer to unpublished notes and papers by Paul Richards on which they base their outline of the technographic approach.

These considerations and assumptions have their implications for the process of data collection in this research, and especially for the validity of this research and its findings. Most importantly, they highlight the importance of selecting interview respondents and other sources for data collection that will allow a diverse perspective on the project under study, and that will allow triangulation of the research findings. Concretely, this has led to a process of selection of interview respondents that included representatives from the projects from different levels in the organisation. Both project managers, as well as employees on 'lower' levels of management and execution of the project were interviewed. Similarly, both stakeholders with a scientific perspective, as well as research managers have been interviewed, representing entirely different perspectives on the dynamics within each of the projects. In addition, next to respondents from the projects under study, respondents have been selected that had a more distanced view on the project. These respondents included people that no longer are part of the project under study, that work for other similar projects, or even that work for NGOs that are highly critical about the specific project under study. These different respondents provided a very rich resource of perspectives on the projects histories, their rationales and the degree to which they are reaching their own internal goals, and may be meeting expectations that other stakeholders have with respect to these projects. The same applies to the various policy documents, newsletters, websites, and other documents that have served as an input for the technographic research underlying the three case studies.

Finally, considering the somewhat contested nature of the projects studied in this research, and the potentially conflicting viewpoints from both within the project organisations, as well as from outsiders, all respondents have been interviewed on the condition of anonymity. This may restrict the transparency of the research design by preventing the publication of a list of interview respondents, but it has encouraged respondents to speak freely about their views and experiences with the projects studied. This is considered to have increased the quality of the case studies to an extent that legitimizes the anonymity of the interviewed respondents.

Research planning

The 4 year research project has been divided in a number of phases, allowing for a more general orientation of the research and the development of a conceptual framework at the beginning, and the consecutive execution of the three main case studies. Concretely, the following phases and activities can be distinguished:

As becomes clear from Table 2.1, the data collection for the case studies has primarily relied on semi-structured interviews with key stakeholders and a wide range of informants for the different case studies. Transcripts of these interviews were sent back to interviewed stakeholders for approbation. In addition a wide range of written documents, policy papers and websites have been collected and have served as inputs for the different case studies. Finally, visits to the different institutes coordinating these projects have been conducted, in

Table 2.1. Main research phases and research activities.

Phase	Period	Main activities
1	November 2004 – April 2006	Literature review on agricultural development in a contemporary and historical perspective Literature review on theories of technology Development of a conceptual framework informing the empirical approach Development of a research design Explorative 1 month stay in India in December 2005 to obtain familiarity with biotechnology innovation in a development context, and the practical operationalization of participatory methodologies in the context of agricultural innovation Series of 7 semi-structured interviews with informants in Indian agricultural and biotechnology development, mainly in New Delhi and Hyderabad Series of 28 semi-structured interviews and informal conversations with informants in the Dutch plant biotechnology and development sector.
2	May 2006 – March 2007	Preparation, execution and analysis of CIMBAA case study (Chapter 4) One month stay in India in December 2006 for a series of 29 interviews with Indian stakeholders from the CIMBAA project, as well as representatives from other organisations in the sector of agricultural development; mainly in New Delhi and Hyderabad
3	April 2007 – September 2007	Preparation, execution and analysis of CIP case study (Chapter 5) One month stay in Peru in June 2007 for a series of 26 interviews with representatives from CIP and other Peruvian organisations in the sector of agricultural development; mainly in Lima and Huancayo.
4	October 2007 – March 2008	Preparation, execution and analysis of GCP case study (Chapter 6) One week visit to Mexico for a series of 6 interviews with stakeholders from the Generation Challenge Programme[1]
5	April 2008 – November 2008	Analysis and comparison of the main case studies and development of the findings into a PhD thesis

[1] The rather limited number of interviews conducted for the GCP case study, in comparison to the other two case studies is explained by the much more focused research questions underlying this case study, by the availability of a great number of highly informative policy documents providing a very rich source of information, and by the generally more upstream nature of this project, limiting the number of other stakeholders that are directly involved in this project or opinionated regarding its work.

order to get an insight in the institutional context in which these projects are carried out and to increase the validity of the obtained research data.

Validity of the study

Although the study presented in this thesis is an explorative one, it should keep an eye on the validity of its findings. Various types of validity are commonly described in the context of applied sociological research. According to Bickman *et al.* (1998), these include:

- Internal validity: the extent to which causal conclusions can be draw from research findings.
- External validity: the extent to which it is possible to generalize from the data and context of the research study to broader populations and settings.
- Construct validity: the extent to which the constructs in the conceptual framework are successfully operationalized in the study.
- Statistical conclusion validity: the extent to which the study has used appropriate design and statistical methods to enable it to detect the effects that are present.

In addition, Bickman *et al.* (1998) note that the relative emphasis on these types of validity may vary, depending on the type of study. For an explorative descriptive study such as this one, they suggest emphasis to be placed on external and construct validity.

Statistical conclusion validity has limited relevance in the context of this study considering the qualitative and exploratory nature of the study and the objective to present illustrative examples of contemporary understanding of 'appropriateness' in technology development projects. This means that the statistical foundation for quantitative research findings is of little relevance. Having said that, the study does aim to maximize its (statistical) conclusion validity by adopting an appropriate research method based upon thick and rich case study descriptions, allowing the researcher to detect and describe the phenomena of interest.

Internal validity is considered to be of limited relevance as well, in the context of descriptive studies (Bickman *et al.* 1998). Rather than providing strong arguments for causal relationships underlying the observed phenomena, the description of the phenomena as such is of importance. Having said that, the study does focus strongly on the context in which these phenomena emerge, and what potential relations could be between this context and the observed phenomena. Moreover, the selection of very different interview respondents (as described in the section on data collection), and the visits to project sites allow for triangulation of research findings, and prevents a bias towards the 'formal version' of a project history. These elements have been employed to maximize the internal validity of this research.

External validity is of importance for descriptive studies that aim to contribute to the wider understanding of a given phenomenon, based upon the study of a specific subset of the population under study. Given the exploratory nature of this research, and the ambition to bring to the surface different ways of dealing with 'appropriateness' in a technology development project, the external validity is limited. The specific interest in case studies that represent different approaches to agro-technological development implies that they do not necessarily provide an average sample of contemporary pro-poor biotechnology development or plant breeding. For the same reason, not all activities of the programmes and institutes studied are given equal attention in the case study descriptions. Instead, some interesting initiatives and aspects of these case studies will be brought to the fore, illustrating different ways in which biotechnology and plant breeding can be made 'appropriate' for resource poor farmers in different contexts. The concrete implication of the methodological approach taken is that this picture cannot be extended to make statements regarding the general state of affairs in contemporary pro-poor biotechnology development, or the pervasiveness of trends of modernisation or industrialisation. In fact, any research which is based upon case study material should be modest in making sweeping conclusions regarding historical trends of industrialisation or modernisation of agro-technological development. However, the approach adopted in this thesis will allow the research to explore the characteristics of a limited set of examples and to distil conceptual lessons from these examples regarding the different dimensions in which technological projects can be made more or less appropriate for resource poor farmers and their problems in agricultural production. As such, in spite of its limitations in terms of external validity, the chosen methodology does fit the goals of this research.

Finally, the construct validity of this research, related to the operationalization (or measurement) of the constructs in the conceptual framework is of upmost importance for this study. This kind of validity deals with the coherence between observed phenomena and the conceptual conclusions that are drawn from these observations. As such, the degree of construct validity of any research is strongly dependent upon the formulation of a clear conceptual framework, and the qualitatively good execution of interviews and other means of data collection. For this study, construct validity has been maximized by the careful formulation of research questions that operationalize the objectives of this research. Moreover, Chapter 3 presents a discussion of the conceptual backgrounds of this study, and provides a detailed discussion of the key concepts that underpin the case study analysis in Chapters 4 to 6. Finally, care has been taken to adopt appropriate interview techniques and coding techniques for the analysis of the research data that prevent a distorted or biased collection or interpretation of research data. Interview reports were analysed by theoretical coding (Flick 2006) based upon a set of sensitizing concepts (Bowen 2006)[23] derived from the literature review and conceptual

[23] Bowen (2006) discusses the use of sensitizing concepts in the context of Grounded Theory as research methodology (Glaser and Strauss 1967), and argues that they are generally used by sociologists as interpretative devices and as a starting point for a qualitative study. Moreover, he quotes Blumer (1954, p. 7) in saying that they give *"the user a general sense of reference and guidance in approaching empirical instances. Whereas definitive concepts provide prescriptions of what to see, sensitizing concepts merely suggest directions along which to look"*

discussions presented in Chapter 3, and which further evolved based upon the key concepts emerging from the interview transcripts.

Concluding remarks

A methodology is only as good as its coherence with the research objective and the research questions asked. The aim of this study as formulated at the beginning of this chapter is clearly explorative. This is reflected by the research questions asked, by the case study selection and by the approach to data collection here described. Implicitly, the research design is influenced by a working hypothesis that the operationalization of 'appropriate technology development' is not straightforward, and that different projects, in different contexts, with different institutional settings and rationales find different ways of legitimizing their work, and in positioning their contribution to agricultural development vis-à-vis other actors involved in this innovation process. In order to bring such a diversity to the surface, and to get an idea of the dimensions in which approaches vary, an explorative research design is required. As has been discussed in the previous section, such a research design does raise questions and introduces limitations in terms of external validity of the study. Nonetheless, it is considered to be the most appropriate way of bringing an expected diversity and creativity in approaches of different projects to the surface. The kind of argument that a study like this can make will not be in terms of generalizing its findings to all other contemporary projects of agro-technology development. Instead, the case study findings can support an argument regarding the differences in approaches observed, their relationship with a wider context, and the repercussions for development policy if these approaches invite new ways of executing or funding the process of innovation.

Chapter 3
Biotechnologies and the transformation of agricultural production systems

"Unable to effectively represent the profusion and complexity of real farms and real fields, high-modernist agriculture has often succeeded in radically simplifying those farms and fields so they can be more directly apprehended, controlled, and managed."

(Scott 1998, p. 262)

"... the commodity form is an underlying and constitutive regularity which shapes and limits the particular forms taken by the episodic and often chaotic expressions of a developing capitalism. To extend the reach of the commodity form is to extend the reach of capitalism. No matter how immense it may already be, the very essence of capitalism is the enlargement of the collection of commodities by which we are already surrounded"

(Kloppenburg 2004, p. 315)

"...some products, such as seeds, can no longer be considered as being just material goods. They create primarily new social relations."

(Ruivenkamp 2005, p. 14)

Introduction – Historical trends in agricultural modernisation and industrialisation [24]

Agricultural development does not take place in a historical vacuum. Instead, it takes place against the context of pervasive historical trends and ideologies in terms of how food production is being organised, and what the role of agricultural producers in that system of production is. The comparison of the historical and socio-political backgrounds of the Green Revolution and the Gene Revolution in the introductory chapter already indicated that the context in

[24] This chapter builds upon conceptual explorations that have been published in an earlier phase as: Vroom, W., G. Ruivenkamp and J. Jongerden (2007). 'Articulating Alternatives: Biotechnology and genomics development within a critical constructivist framework'. *Graduate Journal of Social Science* **4**: 11-33

which agricultural technologies develop has changed, and that this has repercussions for the underlying motivations of technology development and agricultural research. Ignoring such wider historical contexts would imply a risk of making an analysis of contemporary agro-technology development naïve and short-sighted. Therefore, this chapter will delve into some of the literature on such historical trends and ideologies in agro-technological development, and will thereby provide a background against which specific case studies in consecutive chapters can be studied.

This chapter distinguishes between two major sources in literature that provide a background for contemporary agro-technological development. The first is related to the notion of agricultural modernisation and a concern that it has been biased towards a rather homogeneous model of modern agricultural production. As argued in the introductory chapter, a specific Eurocentric modernisation philosophy was present as undercurrent during the Green Revolution, dictating a specific transformation of third world agriculture into modern and industrialized agriculture. While current international agricultural modernisation may be less determined by a converging image of the future of agriculture, this background still raises questions regarding the dynamics in contemporary modernisation processes. Most importantly, a tension arises between modernisation as an imposed condition, prescribing a necessary transformation from 'old-fashioned' production models to modern production systems, and – on the other hand – a more open-ended approach to modernisation that tries to connect more carefully with traditional farming and seed production systems, and aims at a hybridization of modern and traditional farming styles.

Secondly, next to a specific modernisation ideology, this chapter will elaborate on the crucial role of the private sector in contemporary agricultural development. As suggested in the introductory chapter (in reference to Govindan Parayil and others), the agro-food system is not committed to producing sufficient good-quality food alone, it is also organised to serve the commercial interests of a limited set of seed and biotechnology companies, and food manufacturers. The dynamics of an extending capitalist system into agriculture is an important background trend to elaborate, in order to better understand and appreciate current projects of agricultural development, and the way they relate to this trend. As a corollary to this trend, the emergence and extension of a global system for intellectual property protection will be discussed, as well as initiatives to challenge it.

There is a third element, which is necessary to link the discussion on wider trends in agricultural development, to contemporary cases of agro-biotechnology development for resource poor farmers. That element is the link between a specific type of modernity, and technological design. The last part of this chapter will provide a discussion of how technologies are thought to be linked to wider social norms and ideas about modernity through their material design. This will provide an entry point to discuss the cases in the following chapters, with a critical perspective on agricultural modernity.

Agricultural modernisation – From an imposed condition to an open-ended approach

It is tempting and attractive to see agricultural development as a process in which resource poor farmers in developing countries are engaged in the development of technological means to improve their production, increase their income, and improve their livelihoods in a variety of ways because of the improved economic basis under their existence. Development as freedom, so to say (Sen 1999). However, while such optimistic visions certainly circulate in the mainstream discourse on agricultural development, a wide range of scholars has painted an entirely different picture of the dynamics that have characterised processes of agricultural development and modernisation in the past. Rather than a liberating affair in which new agricultural technologies provided an emancipatory potential for the most poor to improve their lives, agricultural development has been unmasked as a sometimes destructive affair in which a wide diversity of local approaches to agricultural production was replaced by an externally imposed, homogeneous model of agricultural production. Sometimes, this model of externally planned agricultural modernisation has been associated with the interests of foreign governments, multinational companies, or ambitious and short-sighted agronomists. However, more in general an ideology of a planned and globally converging agricultural modernity – regardless of direct commercial or political interests – has been mentioned as a structuring force in much agricultural development.

The understanding of the Green Revolution as related to geopolitical interests during the Cold War, and aiming to reduce the spread of communism in the Third World – as described in Chapter 1 – was a first example of such an analysis. Rather than taking the diversity in agricultural production systems as a starting point, and empowering the poorest farmers with new technologies and methodologies, the project was committed to the widespread adoption of a relatively homogeneous model of agricultural production, in order to increase national food production levels. It cannot be denied that this approach has been successful in terms of raising productivity in many places in which the Green Revolution was rolled out. However, it is time to shift focus and to question whether a more diversified approach to agricultural development is possible, and how the emancipatory function of agricultural technologies for resource poor farmers can be taken as a starting point. This would arguably be a more appropriate approach to reach the areas where the Green Revolution so far has failed to have a positive impact on productivity and economic growth.

Other scholars on agricultural modernisation

In order to widen the perspective on agricultural modernisation, it is helpful to review the work of a number of scholars who share a concern over the nature of agricultural modernity, and the way in which policy makers and technologists have advocated a homogeneous picture of agricultural development.

For example, Jan Douwe van der Ploeg has criticized the role of governmental expert systems in recent and contemporary processes of Dutch and European agricultural modernisation. He argues that policy makers have created an image of a 'virtual farmer' which features in their policy documents, but which only represents a very limited fraction of all farmers and their practices (Van der Ploeg [1999] 2003). Van der Ploeg has noticed how very different 'farming styles' can exist next to one another, and how different farmers (with different characters and inclinations) can find very different solutions to socio-economic conditions of production. Van der Ploeg calls farmers that have found a successful mode of production different from the dominant model 'black swans', and claims that in practice they remain largely invisible to policy makers (*ibid.*). In conclusion to his analysis, he argues that agricultural development should be an 'endogenous process' in which different solutions as part of different farming styles can play a role and be exchanged among farmers (See e.g. Van der Ploeg and Long 1994). In addition, in his more recent book 'The New Peasantries', he specifically explores the ongoing significance of peasant modes of production in the face of international agricultural modernisation. He defines the peasant condition as *"the ongoing struggle for autonomy and progress in a context characterised by multiple patterns of dependency and associated processes of exploitation and marginalization"* (Van der Ploeg 2008, p. xiv). In summary, he believes in the creative potential of farmers themselves, and warns for an overly restrictive modernisation policy, which may wipe out any alternative solution to the dominant perspective.

James Scott also rises concern about the simplification typical of high-modernist scientific agriculture (Scott 1998). He elaborates how agriculture has always been about the simplification of natural processes, but reached a peak in 20th century agriculture which was strongly dominated by mechanisation. He quotes Jack Kloppenburg in saying that *"genetic variability is the enemy of mechanization"* (Kloppenburg 1988, p. 177; quoted in Scott 1998, p. 267). This led to the breeding of crop varieties that were highly adapted to mechanical harvesting, through their plant morphology, and uniform fruit size. Not only has this led to rather ridiculous situations in which the uniformity of e.g. tomatoes and their suitability for mechanised harvesting was considered more important than their taste or nutritional value. It also led to agricultural production systems that (according to Scott) are genetically so uniform that they have become extremely vulnerable to disease and pests, and hence the need for blanket pesticide applications to protect crops from such 'biotic stress factors'.

While Scott acknowledges that this model of agricultural modernisation has been successful in the industrializing West, he argues that it has been highly inappropriate for much agricultural production in the developing world. Scott argues that the model of American agricultural modernism was actively exported during the middle of the 20th century. It was strongly committed to the superior technical efficiency of large-scale farms, the importance of mechanisation to save labour, the superiority of monocropping and hybrids over polycropping and landraces, and the advantages of high-input agriculture, including fertilizers and pesticides. However, in spite of lavish credit subsidies and strong administrative backing, these projects – aimed at scale increase and homogenization of agricultural production – generally failed (see also Johnson and Ruttan 1994). In Scott's view, each failure may have had its own peculiarities, but in general, he argues that the level of abstraction at which these projects were conceived was fatal, rendering them blind to the complexity of agricultural production in tropical circumstances.

He contrasts this model of monocropping and genetic uniformity with the polycropping approach which is for example widespread in Africa. He argues that this model better protects the thin fertile soil against erosion than monocropping systems, and is better adapted to deal with the variable timing of rains in tropical climates. Moreover, it functions as a risk avoiding system, since different crops that are cultivated at the same time perform better under different circumstances, and provide food at different times during the year. In addition, the genetic diversity in such polycropping systems makes them less vulnerable to diseases and pests. However, as Scott argues, from a high-modernist point of view, this type of agriculture simply looks disorderly. As he puts it: *"The high-modernist aesthetic and ideology of most colonial agronomists and their Western-trained successors foreclosed a dispassionate examination of local cultivation practices, which were regarded as deplorable customs for which modern, scientific farming was the corrective."* (Scott 1998, p. 279) Moreover, standardized external agricultural advice fitted badly with the unavoidable variation by farm and fields which *"requires a more open-ended approach, with, in all probability, farmers doing much of the necessary experimentation for themselves"* (Richards 1985, p. 61; quoted in Scott 1998, p. 284). In other words, agricultural development strongly focused on the import of a homogeneous model of farming, and was relatively blind to the advantages of already existing polycropping systems, and the sophistication of the knowledge and experience of local farmers.

So, while Van der Ploeg focused on the way in which agricultural policy supported a uniform image of agricultural production and modernisation, Scott translates this to technical systems and infrastructures, and notes how for example the Green Revolution has been an example of a 'simplification' of agricultural production that made sense from the perspective of an external observer, but had very little connection with the reality and value of diverse agricultural production systems in developing countries. Again, development did not start with a thorough analysis of the existing situation and the emancipatory potential of new technologies. Instead, it rolled out a homogeneous picture of high-modernist production.

A third example of how agricultural modernisation raises questions regarding its relationship with existing production systems relates to the development of seed systems. A seed system can be defined as *"all the elements of seed provision that interact as a system, including for example genetic resource management and crop improvement, multiplication and diffusion/marketing"* (Louwaars 2007, p. 149). Niels Louwaars has elaborated how many developing country governments have developed seed policies and regulation to guide the further evolution of the seed sector in their countries. Louwaars argues that these policies were highly influenced by a seed system development paradigm that was published by Douglas in

1980, and which laid down a linear approach for the transformation of current informal seed system into what would be a modern, sustainable, commercial seed system (Douglas 1980). This approach for seed system development largely ignored the value and intricacies of already existing informal (or farmers') seed systems. In response, Louwaars argued that informal seed systems still have a crucial role to play in the seed provision for many resource poor farmers in developing countries. He therefore developed an alternative framework for the development of seed systems, focused on the integration of formal and informal seed systems, rather than on the replacement of informal farmers' seed systems by formal seed systems. In practice, such an integrated 'diversified seed system' would depend on a range of interactions between formal- and informal seed systems, on the level of the collection and management of genetic resources, (participatory) plant breeding, seed multiplication and marketing. On each of these levels, formal- and informal seed systems have different, and potentially complementary assets and capabilities, that – according to Louwaars – can be used optimally when combined (Louwaars 1994; Louwaars and Almekinders 2002; Louwaars 2007).

Implications: limits to the validity of modernisation

What these scholars have in common is that they focus on the transformative nature of agricultural modernisation, and raise the concern that this process has little consideration for the values of traditional cropping systems, has little consideration for diversity in production systems and potential solutions to problems of productivity, and has little concern for the role of farmers/peasants themselves in this transformation process. Instead, agricultural modernisation is to an important extent presented as an imposed condition rather than a liberating one, and – secondly – as a process that is destructive of traditional production styles and seed systems, instead of linking up with them and making use of the already available local knowledge and experience. In referring to similar trends, Joost Jongerden has argued that in the course of 20th century agricultural modernisation, traditional 'peasants' were transformed into modern, entrepreneurial 'farmers' (Jongerden 2008). He too notes how this process was hardly about the combination of useful elements of new scientific knowledge and local experience, but rather about the replacement of one production model by another. To him, this type of agricultural modernisation is a process of 'creative destruction', in which *"The destruction of the peasantry was productive in the sense that it was contingent on the creation of a new class of farmers. In other words, the death of the peasantry is associated to the birth of the farmer, a modern entrepreneur, integrated into agro-industrial chains and producing primarily for market consumption."* (Jongerden 2008, p. 125)[25]

[25] Multiple sources are mentioned for the notion of creative destruction, but it is clear that it was popularized by Joseph Shumpeter when he used the term in 1942 to describe a process of transformation that involved radical innovation. He considered it to be *"the essential fact about capitalism. It is what capitalism consist in and what every capitalist concern has got to live in"*(Shumpeter [1942] 1975, p. 84). Jongerden also refers to Walter Benjamin, who described modernity as *"inconceivable without its destructive, cathartic side: the liquidation of the value of tradition"* in his 'Work of art in the age of reproducibility', which was first published in 1936 (Benjamin 2003, p. 254; quoted in Jongerden 2008, p. 125).

Such an understanding of agricultural development raises concern. James Scott argues that the simplification of high-modernist schemes to improve the human condition commonly fails, because it fails to take into account the complexities and variability of real life production systems. Niels Louwaars argued that a formal, commercialized seed system may have its value for some farmers, but that it fails to take into account that many resource poor farmers do not obtain their seed from formal sources, and hence are relying upon informal seed systems. Here too, formalization has its limits. And Jan Douwe van der Ploeg argued that agricultural policies are increasingly concerned with a virtual farmer that does not exist. He urges policy makers to take diversity in farming styles seriously and to take it as starting point for the making of agricultural policy. Finally, Joost Jongerden draws attention to the fact that agricultural modernisation has had very destructive effects on farmer communities which were supposed to modernize or disappear. So, while a uniform picture of farming and agricultural modernisation is capable of increasing productivity for some farmers, it does reach the limits of its validity in the situations described by these authors.

This provides an important background for the case study analysis in this thesis. It brings into focus the way in which agricultural development is able to link up with existing production styles and seed systems, and how it is managing to bring the perspectives and interests of local stakeholders on board in the process of development.

Industrialisation of agriculture – Issues of control in the Third Agro-Food order

As Govindan Parayil argued in Chapter 1, the dynamics of the Gene Revolution are to an important extent determined by the interests of private sector companies, that see new crop varieties not primarily as a way to alleviate hunger and poverty, but as a way to enter new markets and to maximize shareholder profits (Parayil 2003). This understanding of the current agro-food system is confirmed by Pistorius and Van Wijk who have described three main agro-food orders that are characterised by a different division of labour between the public and private sector, by different regulatory frameworks protecting innovations, and by different dominant crop development policies (Pistorius and Van Wijk 1999) (Table 3.1). In their categorization, the Third Agro-Food Order started in the 1980s and is characterised by an increasing importance of genetic technologies in plant breeding, decreasing government funding of plant research, and an increasingly important role of the private sector in plant breeding. In addition, Pistorius and Van Wijk distinguish between a set of three rival agricultural production strategies: market-led agro-industrialisation, state-led agro-industrialisation and a third strategy directed to non-industrial, farmer-oriented agricultural production (Pistorius and Van Wijk 1999, p. 20). These agricultural production strategies are no more than ideal types, but provide an illustrative framework to discuss different approaches to agricultural development around the world, and through the years.

A crucial question that emerges is why different agricultural production strategies emerge in different countries and different historical epochs. Pistorius and Van Wijk argue that there is an important relationship between the preferred agricultural production strategy, and the agro-food order that a specific country is in (see also Table 3.1). In summary, they argue that state-led industrialisation of agriculture was the dominant agricultural production strategy during the second agro-food order, ranging approximately from the 1930s to the 1980s. Market led industrialisation of agriculture would become dominant in the third agro-food order, starting in the 1980s. Remarkably, the non-industrial farmer-oriented agricultural production strategy has not been a dominant agricultural production strategy in any of the agro-food orders. However, Pistorius and Van Wijk argue that it has become a relevant strategy with the introduction of the Green Revolution, when in developing countries a large part of the rural population was marginalized, since particularly the peasantry could not be incorporated in the process of agro-industrialisation.

This section will address both the changing division of labour in the current agro-food order as part of an industrialisation of agriculture, as well as the accordingly changing regulatory

Table 3.1. Crop development policies in three historical Agro-Food Orders. (Adapted from Pistorius and Van Wijk (1999), p. 24.).

Agro-food order	International division of labour in agriculture	Dominant agricultural production strategy	Aim and features of dominant crop development policy
First: 1870s-1930s	Settler states, notably USA, export grain to feed urban labour in European metropolises	Initial state intervention to support national temperate agricultural sector in industrial core countries	Improving ompetitiveness of national temperate agricultural sector
Second: 1930s-1980s	North America and Europe become food exporters. Developing countries emerge on world markets as importers of temperate cereals	State-led industrialisation of agriculture (Exported to and adopted by developing countries in 'Green Revolution')	Rise of productivity in staple crops
Third: 1980s-	OECD countries strengthen position as major food exporters. Developing countries obtain niches in world market	Market-led industrialisation of agriculture	Rapid adaptation of plant qualities to diverse and changing world markets

structures surrounding it. First, the main organizing principles of the process of agro-industrialisation will be discussed. This is relevant, since the capital accumulation process within the context of the agro-food system is generally recognized as being different from non-food systems. James Kirwan notes three main reasons: first because it is dependant upon an inflexible land base, secondly because it is heavily dependent upon organic properties throughout the linkages from production to consumption, and thirdly because the demand for food is relatively inelastic, particularly in Western Economies (Whatmore 1995; Kirwan 2003). The result of these constraints is that capital has traditionally sought to reduce the impact of 'nature' through the application of science and technology, and secondly to maintain market growth through adding value to agricultural products outside the immediate production-consumption cycle. Already in 1987, Goodman *et al.* have described how this is achieved through mechanisms of appropriationism and substitutionism, which serve as organizing principles of agricultural industrialisation (Goodman *et al.* 1987).

Appropriationism

Appropriationism refers to a process in which elements of farming practice are gradually taken over by external institutions, are transformed into industrial activities and then re-incorporated into agriculture as inputs. This process for example describes the externalization of seed breeding and multiplication. While traditionally farmers have been the key protagonists in varietal management and experimental breeding, this element of agricultural production has increasingly been taken over by specialized seed breeding companies. Good quality seed was transformed from an on-farm product into an agricultural input to be obtained externally. Similarly, soil fertility management and pest insect management have been externalized by the introduction of chemical fertilizers and pesticides. Integral elements of farming practice have been 'appropriated' and transformed into industrial inputs.

Appropriationism is an important tool in minimizing the impact of 'nature' on agricultural production systems by industrializing its different components. In fact, appropriationism is argued to be an act of disconnecting agricultural production from its natural environment, allowing a standardization of agricultural production (Ruivenkamp 1989, 2003a). This leads to discussions over the sustainability of agricultural production. First of all, the standardization of agricultural production, and the reduction of the number of cultivated varieties to a limited number of commercially available varieties has been argued to have led to a strong erosion in agricultural biodiversity. While the precise effects of the narrowing of the genetic base under agricultural production are not entirely clear or uncontested (see e.g. Brush 1992; Smale 1997), a too narrow range of genetic variety is feared to make agricultural production systems vulnerable to newly emerging diseases and pests. In addition, the disconnection between agricultural production and the natural environment is argued to be a reason for the unsustainability of the current conventional production methods. The chemical control of soil fertility and pest infestation is leading to the pollution of the natural environment (Pretty 2002).

Next to the questions regarding sustainability of agricultural production, raised by appropriation, the process has important consequences for the social roles and responsibilities in agricultural innovation and production. It is argued to effectively lead to a deskilling of farmers and an increased dependence upon external institutions (Ruivenkamp 2003a). The latter is beautifully illustrated by the seminal study of the political economy of plant biotechnology by Jack Kloppenburg (Kloppenburg 1988), and more specifically his study of the development of hybrid maize in the United States. Kloppenburg described that, like other cereals, maize was conventionally grown by farmers by replanting a portion of their harvest. This way, maize had a dual function as both grain for consumption and marketing, and seed for the next growing season. This changed with the introduction of hybrid seeds. Hybrids are derived from a cross with two genetically homozygous parental lines, and may benefit from 'heterosis,' or hybrid vigour, that occurs when different parents produce strongly heterozygous hybrid offspring. Heterosis is claimed to significantly increase yields as compared to non-hybrid varieties.[26] Essential for the marketing of hybrid seed is that the generation that is marketed is genetically heterozygous. Although technically fertile, these hybrids create a strong segregation of agronomically valuable traits in a next generation, rendering their offspring much less attractive for commercial cultivation. In practice, this has meant that farmers are strongly encouraged to buy new seed every year, instead of reusing their farm saved seed for several years like they have done traditionally.

Importantly, the industrial appropriation of elements of agricultural production goes hand in hand with questions of control. Like mentioned before, it is about minimizing the impact of 'nature' on agricultural production. Yet at the same time, it is an important tool in creating a production system in which the supplier of agricultural inputs is an essential element in the production process. The case of hybrid maize seed clearly illustrates this, since hybrids not only increased yields, they also fundamentally changed the agricultural production chain, and firmly embedded public or private external seed suppliers in the line of production.

Taking this argument one step further, Ruivenkamp has argued that appropriationism also opens the door for a far-reaching influence of the seed supplier on the agricultural production system. Next to the annual replacement of seed, this may for example have an influence on other kinds of industrial inputs that are being used. The clearest and most famous example of this is the combined development by Monsanto of the herbicide Roundup®, and Roundup Ready® crop varieties with a transgenic resistance to this herbicide. Selling the seed to farmers

[26] This notion that hybrids are responsible for yield increases may be generally accepted, it is strongly challenged by Kloppenburg who argued that the development of alternatives such as open-pollinated varieties was largely abandoned, and therefore does not provide a fair comparison (Kloppenburg [1988] 2004, p. 92ff).

introduced a strong inclination to buy the herbicide as well (Dutfield 2003).[27] More subtly, assumptions regarding and preferences for specific farming systems are introduced by breeding crop varieties that are suitable for mechanised harvesting, that are specifically evaluated for a monocropping farming systems, or that have short stems and little leafy material, creating a product with little added value for use as animal fodder, or as construction material. In other words, a whole range of assumptions regarding the farming system appears to be embedded in the breeding of specific crop varieties, leading Ruivenkamp to argue that seeds are 'politicising products' that have a crucial structuring function in the development of agriculture (Ruivenkamp 2005). The last sections of this chapter will further delve into the question how – and to what extent – a technological artefact can have a structuring role in the context of application.

Substitutionism

Next to appropriationism, Goodman *et al.* described substitutionism as a second organizing principle in the industrialisation of agriculture (Goodman *et al.* 1987). Substitutionism refers to the process in which agricultural products are reduced to an industrial input, allowing their interchangeability and even their replacement by artificial, non-agricultural components in food manufacturing. In practice this means for example that different agricultural products (sugar cane, sugar beet) can serve as sources of 'carbohydrates', regardless of the precise origin of these components. And in a subsequent stage, such natural sources of sugar can altogether be replaced by synthetic sweeteners like aspartame (see also Ruivenkamp 1986). The replacement of milk by vegetable fats in the production of margarine as alternative to butter is another well known example (Goodman *et al.* 1987).

For substitutionism, the rise of enzyme technology has been very important in the transformation of agricultural products into a limited set of basic ingredients for the food industry, allowing their interchangeability. An important example has been the use of the glucose isomerase enzyme for the transformation of glucose, derived from corn starch, to high fructose corn syrups (HFCS). This low-calorie sweetener emerged as an alternative to sugar from sugar beet or sugar cane, and has become very popular in the food industry (notably the soft drink industry). Goodman *et al.* (1987) note that the competitiveness of the HFCS industry depended crucially on the genetic improvement of glucoamylase enzymes in order to produce HFCS with approximately 42% fructose content. By providing the technical means to produce HFCS with these characteristics, enzyme technology had a crucial restructuring effect on the global sugar production and trade, strongly reducing the sugar imports of the

[27] This example of Roundup Ready crops is merely illustrative for a mechanism in which the seed represents a crucial strategic value. In the case of Roundup Ready crops, it is questionable whether this strategy will work in the long run as the production of the Roundup pesticide is no longer protected by patents, and farmers might as well buy a similar herbicide from another company (Dutfield 2003).

United States, and placing sugar cane production the Phillipines in a tight spot (Goodman *et al.* 1987; Hobbelink 1991).

Substitution of agricultural products by their reduction to mere agro-industrial inputs, essentially requires a disconnection between the agricultural product and its inherent nutritional and culinary value (Ruivenkamp 1989, 2003a). Moreover, it denotes a strong disconnection between consumer and producer, since the agricultural or geographic origin of a final food product is no longer visible, nor relevant. While this is an interesting trend for a globalized food industry, it has raised some questions regarding transparency and traceability of food stuffs throughout the food chain (Opara 2003). The emergence of a number of 'alternative strategies' in food production, focused on the local production of high quality food, and on transparency in the production chain, indicate that consumers are increasingly concerned about the current disconnection between their food and its agricultural origins. Moreover, it is becoming clearer that the supply of high quality food in the richer parts of the world can go hand in hand with hunger and poverty in the places where this food was originally produced. This kind of contradiction in the current agro-food system has led to increasing interest in regional food production and consumption, and fair trade policies aimed at re-establishing a connection between consumers and producers, thereby making consumers aware of the production conditions in which their food was cultivated (Kirwan 2003; Levidow 2008).

Importantly, the reduction of agricultural products to industrial inputs and their interchangeability means that agricultural production can be connected to entirely different industries, other than food production. This element of substitutionism is instrumental in the adding of value to agricultural products, outside of the immediate production-consumption cycle, which was mentioned as one of the strategies allowing capitalist accumulation of value in agriculture. The most obvious example is the recent use of agricultural products as sources of carbohydrates and oils as inputs for the emerging biofuels industry. Here, the importance of enzyme technology once more comes to the fore, since cellulase enzymes can play a key role in the production of bio-ethanol from agricultural waste products (leaf and stem material) (Lin and Tanaka 2006).

This use of agricultural products for different industries (and notably the biofuel industry) can provide opportunities for farmers, since it potentially opens up new markets. At the same time, it introduces a competition over agricultural products for food and fuel, which is feared to have strong correlations with fluctuating and growing food prices. For example, in 2007 world corn prices reached a peak, and it is generally assumed that this was in part caused by the increased interest in corn as source for bio-ethanol. This meant that prices of corn as food product became prohibitively high for consumers in Mexico (Ford Runge and Senauer 2007). Whether this effect can be completely blamed on the dual use of corn for both food and fuel remains open to question. Moreover, there might be ways of using biomass as a fuel, without creating a fierce competition over agricultural products between food and fuel.[28] Nonetheless, it is clear that biotechnology mediated processes of substitution and interchangeability can have important restructuring effects on the international production and trade of agricultural products, and on the strong position of food and fuel manufacturing companies that benefit the most from the interchangeability of their inputs for industrial production.

Relevance for agricultural development

The key argument of Goodman *et al.* (1987) was that agricultural industrialisation was characterised by these two organizing principles: appropriationism and substitutionism. These processes are essentially about the distribution of power and commercial benefits in the agro-production system.[29] They may be legitimized by seemingly objective and technical advances in production technology, but they also bring about a social reorganisation of the agro-industrial production chain (Ruivenkamp 1989, 2005). That is not to say that these processes are inherently bad or inappropriate for developing countries. In fact, a certain degree of industrialisation and commercialization of agriculture may be a good way of increasing national food production in some countries. As Reardon and Barrett (2000) write, it is very difficult to precisely evaluate the welfare impacts of a process like agro-industrialisation. While benefits of industrialisation may be differentiated and of most direct interest to medium- and large scale farmers, economic growth may contribute to increased rural employment and the availability of food for lower prices, increasing real wages, and contributing to progressive growth.

But the discussion of appropriationism and substitutionism does raise the question whether other dynamics in agricultural development are possible, and whether they may counteract the trends of homogenization and externalization in agricultural production that are typical for the industrialisation of agriculture. Is contemporary agricultural development for resource poor farmers dominated by the here presented model of agricultural industrialisation, or can a wider diversity in approaches be witnessed? Do such projects not only consider the appropriateness of technical solutions in technical terms, but also in terms of the different roles of seed suppliers, farmers, and food manufacturing companies in the agricultural innovation and production system? And if so, how is that translated into the technologies that these projects develop? These are the questions that will be taken on board in the case study analysis in Chapters 4 to 6.

[28] For example through the use of agricultural waste products, or through the use of crops that do not compete with food production, like Jatropha. These approaches to the production of biofuels is generally indicated as 'second generation' biofuels.

[29] In reference to this issue of the distribution of power, Ruivenkamp speaks of an increasing 'control from a distance' by life science companies and research institutes, over agricultural production (Ruivenkamp 2003a).

Intellectual property – The crucial importance of ownership

A corollary to the industrialisation of agriculture and the increasingly important role of the private sector in the production of agricultural inputs, is the development of a global system for intellectual property (IP) that protects the investments in R&D by these companies. Jack Kloppenburg already noted how *"Capital has pursued two distinct but intersecting routes to the commodification of the seed. One route is technical in nature and the other social. [...] The seed can be rendered a commodity by legislative fiat as well as technical force."* (Kloppenburg [1988] 2004, p. 11) Kloppenburg here refers to (1) the development of hybrid corn varieties, turning seed that was previously freely exchanged into a commodity; and (2) to the passage of new legislation on the protection of breeders' rights, prohibiting the multiplication of seed by others than the official breeder who is holding the ownership to a specific plant variety. This section will delve into the second route of the commodification of seed as industrial input.

The extension of the intellectual property system

A wide range of scholars has described how the protection of intellectual property in biotechnology and plant breeding has undergone important changes over the last three decades (Falcon and Fowler 2002; Brush 2003; Dutfield 2003; Gepts 2004; Adi 2006). From a situation – pre World War II – in which much of the basic and applied seed technology for agriculture originated as public goods from the public sector, the involvement of the private sector has become increasingly important and has to an important extent been determining the political economy in plant breeding (Kloppenburg 1988). This involvement of the private sector in plant breeding has been accompanied by the rise of an increasingly strict intellectual property regime, which was both required by, and stimulated the privatization of the seed business.

Since the late 1960s, the investments of plant breeders in new varieties have been protected by Plant Variety Protection (PVP), which first emerged in the US, and has since been internationally established in several UPOV conventions in the past decades (Le Buanec 2004).[30] The requirement for this type of protection is that plant varieties be sufficiently 'distinct, uniform and stable', and it grants the breeder the right to exclusive commercialization for a period of 17 years. However, there is nothing in PVPs (or Plant Breeders Rights, PBRs) that stops others from using the same material for further crosses (commonly known as the *'breeders' exemption'*), or stops farmers from saving, exchanging and replanting their own produced seed (commonly known as *'farmers' rights'*).[31]

[30] UPOV is a French acronym for "Union pour la Protection des Obtentions Végétales"; in English: The International Union for the Protection of New Varieties of Plants.

[31] It should be noted that the UPOV convention has been revised a number of times. The most recent 1991 revision of UPOV has restricted the farmers' privilege by limiting the commercial multiplication and sale of farm saved seed, although some controversy remains of the exact extent of the limits that have been put on farmers' to re-use their own generated seed. Compare e.g. Adi (2006, p. 104) with Smolders (2005, p. 4), or Dutfield (2003, p. 189).

At this time, the patent system – which has more recently been causing so much debate in biotechnology – had little relevance in the context of agriculture. Plant patents have existed in the US since 1930, but only provided protection for clonally propagated crops (such as roses or apples) and were therefore not widely applied to agricultural crops, nor was the plant-patent system extended widely beyond the US (Dutfield 2003, p. 181-184; Le Buanec 2004, p. 2-3). This changed with the emergence of more strict systems of intellectual property protection starting from the beginning of the 1980s, notably with the 1980 *Diamond vs. Chakrabarty* case in which the US Supreme Court ruled that a living micro-organism, constructed by gene-transfer technology was patentable (Chiarolla 2006, p. 32). This opened the door to a trend among biotechnology companies to maximise their number of biotechnology patents, as rapidly as possible (Falcon and Fowler 2002). This has resulted in the intellectual property protection of genes, traits, molecular constructs and transformation procedures; so called enabling-technologies.

This change in patent law implied an important extension of the patent system into the domain of the emerging life sciences. In addition, the use of patents has increasingly permeated into the domain of protection of plant varieties. This can occur for a number of reasons, but most importantly when a 'non-biological' process is claimed for the production of a plant variety[32], or when a patented DNA sequence has been introduced into a plant variety in which it functions (Chiarolla 2006, p. 33).

In addition to an extension of the patent system into the domain of biologicals and plant varieties, various scholars have indicated a geographical extension of the patent system (Dutfield 2003; Gepts 2004; Kloppenburg 2004; Koo *et al.* 2004). Both patents and plant variety rights have geographic boundaries: they need to be filed in a specific country to be valid. If a technology is protected in the US or Europe, this does not restrict the use of this technology in India, as long as the patent is not also filed in that country. Similarly, Plant Variety Protection is only valid in countries that are UPOV members and have installed legislation which enforces UPOV-like plant breeders rights. However, a global harmonization of IP rights is taking place under the influence of the agreement on Trade Related Aspects of Intellectual Property Rights (TRIPS) (Dutfield 2003, p. 191-192; Gepts 2004). Concretely, TRIPS has a specific clause in Article 27(3)b on the protection of plant varieties providing an option to exclude them from patent protection if the country provides for 'an effective *sui generis* system.'[33] It is generally acknowledged that UPOV would constitute such a '*sui generis*' system (Louwaars 2007, p. 96). This means that TRIPS effectively stimulates the global adoption of Plant Variety Protection for plant varieties, and the global adoption of patent law for other biological innovations.

---

[32] This could for example refer to the use of molecular selection methods in plant breeding. See Jansen (1997) for an example of such a patent claim.

[33] *Sui generis* means 'of its own kind', and in practice refers to a system tailored to the specific needs of a country with respect to the specific category of plant variety rights.

The changing intellectual property landscape in biotechnology is generally considered to have had profound consequences for the structure of the biotechnology and seed breeding sector, and vice versa. For example, Pistorius and Van Wijk have described the rise of utility patents in plant breeding, in response to a number of historical developments in the sector. These included the increasing private investment in crop development from the 1980s onwards (meaning there was more economic value to protect), an increasing number of competitors in crop development and the involvement of chemical and pharmaceutical corporations in plant breeding, and an increasing number of collaborative linkages, requiring a more precise demarcation of property rights (Pistorius and Van Wijk 1999, p. 137-138). In addition, Le Buanec mentions how the patent system was attractive for biotechnology companies in order to restrict the possibility for competing companies to use newly released varieties (which is possible under Plant Variety Protection through the *breeders' exemption*), and to limit the use of farm-saved seed (i.e. the *farmers' right*) (Le Buanec 2004, p. 2-3).

At the same time, the growing potential for intellectual property protection in the Life Sciences is argued to have had an important influence on the structure of the biotech industry, and have led to a strong wave of investments in, and acquisition of biotech start-up companies (Falcon and Fowler 2002). In other words, it is not just the private sector that called for a stricter intellectual property protection, it is also the changing legal environment which allowed the private sector to become so dominant in the Life Sciences. In the 1990s this led to a dynamic in which large multinational life science corporations began to buy up small biotech firms and seed business at a very rapid pace, fundamentally changing the structure of the entire sector. The main outcome of this round of reshuffling was a major concentration of the biotech and seed sector, with a very strong position of only a handful of very large life science corporations. Graham Dutfield describes how this led to the emergence of a type of integrated business enterprise called the 'Life Science Corporation', which are often so large that they hold dominant positions in two or more industrial fields that were previously considered to be completely separate. (Dutfield 2003, p. 147).

Challenging the trend towards a limited access to technologies and new varieties

The increasingly strict protection of genetic (enabling) technologies as well as biological material (improved varieties, parental lines) is not without problems or controversies. Both science in general and plant breeding in particular have historically been activities that depended upon the exchange of new ideas and material, increasing the overall level of innovation. With the contemporary sometimes clogged patent landscapes, it becomes increasingly difficult for both private sector companies, as well as public sector institutes to find the necessary 'freedom to operate' (FTO) to do the research they find relevant, and more importantly: to commercialize new findings. A research exemption is often easily negotiated, but the commercialization down the line may be troublesome.

This problem with gaining freedom to operate both exists for private sector companies, as well as for public sector research institutes. However, an important difference is that big multinational life science corporations generally have a much stronger ability to negotiate their way to intellectual property (IP) owned by another company, either by paying for a licence, by the mutual exchange of licences to patented technologies, or by simply acquiring a company with essential IP. In contrast, for public sector institutes these ways of accessing proprietary technologies are generally prohibitively expensive or simply impossible. This is especially raising concerns with respect to the freedom to operate of research institutes that are developing biotechnologies and new plant varieties in or for developing countries (Atkinson *et al.* 2003; Adi 2006).

In response, initiatives for humanitarian use licences have been setup, according to the idea that companies may develop, protect and commercialize certain technologies in developed countries, but can make these technologies freely accessibly or available at low cost for developing countries, in which no substantial commercial market for the proprietary technology exists anyway (Brewster *et al.* 2005) (see Box 3.1 for an example). Next to indicating that such technologies have limited commercial value in less developed countries, such initiatives capitalize upon the potential profit in public relations that may arise for generous companies. One of the important limitations to the concept of humanitarian licences is the transaction costs involved in finding out who relevant patent holders are, and by negotiating the terms of accessing the protected technology. This can in part be solved by adopting standard protocols for accessing a wide range of technologies along a range of institutions. This solution has for example been materialized in the consortium agreement of the Generation Challenge Programme, in which any company or institute joining the research funded by this programme is automatically making newly developed technologies available under a humanitarian licence (Louwaars 2007, p. 129).[34]

As an alternative to humanitarian licences, patent pools can be created. The 'Public sector Intellectual Property Resource for Agriculture' initiative (PIPRA) is an example of such an attempt to pool currently highly fragmented public sector IP in order to create some freedom to operate for the not-for-profit sector to develop pro-poor biotechnologies and crop varieties (Atkinson *et al.* 2003). Their argument is that although public sector institutes (universities) have contributed very significantly to biotechnology development, and (together) even own a large part of all intellectual property on plant biotechnology, their IP is so fragmented across many organisations that no single organisation can use it to create adequate freedom to operate in developing and commercializing biotechnologies for resource poor farmers (Graff *et al.* 2003).

[34] See http://www.generationcp.org/ (last accessed 17 September 2008) or Chapter 6 for more information about the Generation Challenge Programme.

Box 3.1. Humanitarian licences: the case of Golden Rice™.

The most widely quoted example of how humanitarian use licences can provide access to proprietary technology is the case of 'Golden Rice'. The widespread vitamin A deficiency in Africa and South-East Asia provided a motive to develop a rice variety which would contain high levels of beta-carotene, or pro-vitamin A. Swiss researchers Ingo Portrykus and Peter Beyer managed to develop such a rice variety with the help of transgenic technology, transforming the plant with genes from daffodil and a soil bacterium *Erwinia uredovora* (Ye *et al.* 2000). The production of beta-carotene resulted in yellow-orange rice grains, hence the name 'Golden Rice'. However, much of the technology they needed to develop this beta-carotene producing rice was owned by universities and biotechnology companies. In fact, some 70 intellectual property rights from 32 different institutes and companies applied, although the extent to which these patents are actually blocking the development and commercialization differs depending on the country of release (Kryder *et al.* 2000). The gaining of access to all these proprietary elements seemed like a daunting and near-impossible task, and for that reason the ownership of the project was transferred to the multinational seed and biotechnology company Syngenta. This company was able to negotiate humanitarian licences to all crucial elements of the technology. Six key-patent holders were approached[1], and an agreement was reached that allowed Potrykus to grant licences, free of charge, to developing countries, with the right to sub-license (Verbeure *et al.* 2006). Because of this result, the Golden Rice case is widely mentioned as an example of a successful way of dealing with 'patent thickets' (the combination of a large number of patents that apply on a technology). At the same time, the expected positive publicity of this model project is expected to have played a role in the willingness of patent holders to grant humanitarian licences. This means that the case may not be entirely representative as an example of how easy it is to negotiate humanitarian licences on proprietary technologies.

[1] These six key patent holders included Syngenta, Bayer, Monsanto, Orynova BV, and Zeneca Mogen BV, according to http://www.goldenrice.org/Content2-How/how9_IP.html (last accessed 17 September 2008).

In addition, there is a number of 'clearing house' constructions, or technology brokering organisations that aim to make proprietary technology available for developing world agriculture, often by arranging public private partnerships. Such organisations include the International Service for the Acquisition of Agri-biotech Applications (ISAAA), and the African Agricultural Technology Foundation (AATF). Also, the USAID sponsored Agricultural Biotechnology Support Projects (ABSP) are essentially aimed at striking public private partnerships between biotechnology companies and public sector institutes in a wide range of countries, in order to facilitate the transfer of biotechnologies.[35]

Finally, initiatives to experiment with open-source biotechnology have been set up, allowing a greatly facilitated access to currently protected enabling technologies. The idea of open-source innovation and protection has its background in the development of open-source software and has been highly successful in the development of the Linux operating system, and in the building of the Wikipedia online encyclopaedia. It primarily relies on a dynamic in which a great number of small voluntary contributions result in a common resource for all. Cambia is an organisation that has most prominently advanced the notion of open-source biotechnology development. It attempts to develop a protected commons of enabling technologies that allow a 'work-around' for existing (but patented) key technologies, allowing players in developing countries to cheaply use and commercialize new products using these open source technologies (Herrera 2005; Jefferson 2006).[36]

Summarizing, there is a number of initiatives that try to create the freedom to operate necessary to develop and commercialize modern biotechnology, especially for developing countries. What these initiatives have in common is that they work from outside the corporate sphere to increase freedom to operate, by facilitating access to protected technologies, or by creating pools of public sector IP which at least makes the space not occupied by corporate IP much easier to navigate and to use. What these initiatives do not do is to question the technologies themselves, or their appropriateness for developing world agriculture. The basic argument here is that access to (enabling) technologies is a first precondition to start talking about how to develop concrete technologies for resource poor farmers. Once technologies are freely available to a wider range of research institutions or companies, the doors are open to a more contextualized and tailored process of technological design. Whether and to what extent that requires a profound adaptation of the technologies themselves is an important question for the upcoming sections and the case studies in later chapters.

The background provided in this section is of specific relevance to the case study presented in Chapter 4. While there are no indications for a direct relationship between the ownership structure of a new technology and its technical design, the protection of biotechnologies can have important implications for the choices that are being made in terms of what technology is being developed, and in terms of what partners are chosen to collaborate with. Especially in the context of making agro-biotechnologies available for resource poor farmers, it becomes

---

35 See http://www.isaaa.org/, http://www.aatf-africa.org/, or http://www.absp2.cornell.edu/ for more information about these initiatives. (All websites last accessed on 17 September 2008).

36 See Hope (2004), Hughes (2005), or Deibel (2006) for a discussion of 'open-source' in biotechnology. See also: http://www.cambia.org for more information about Cambia's open-source strategy (last accessed 17 September 2008).

crucial who owns certain essential pieces of (enabling) technology, who is capable and willing to negotiate licences, and what partners may be attracted for a specific project in order to unlock potentially useful technology for resource poor farmers. These aspects will return in the case study discussed in Chapter 4, which will present the work of a public private consortium and will address the extent to which intellectual property plays a role in structuring the institutional setup of the consortium. The cases presented in Chapter 5 and 6 primarily take place in a public sector context, and – as will become clear – in those cases the intellectual property framework plays no major role in the analysis.

Recapitulation

The previous sections have elaborated a number of historical trends that provide an important backdrop for the analysis of specific case studies in the following chapters. The first section focused on a specific approach to agricultural modernisation, which was characterised by a creative destructive nature, and was highly externally planned. Interestingly, the discussion on the modernisation of agriculture reaches beyond the concerns about the influence of the private sector in agricultural development, but emphatically includes public sector initiated agricultural development. The key tension it raises is one between modernizing production systems by replacing existing production systems on the one hand, and by linking up with such existing production systems and appreciating their diversity on the other hand. The ability to involve farmer communities in their process of agricultural development, and the ability to hybridize modern and traditional elements in a developing production system, were proposed as important reference points for the case study analysis in future chapters.

The second section elaborated how a historical trend in the industrialisation of agriculture has been characterised by mechanisms of appropriationism and substitutionism, which are mediated and increasingly made possible by the use of biotechnologies. This raised questions on the centrality of the role of the suppliers of agricultural inputs, notably that of seed companies, and on the extent to which a range of assumptions regarding the farming system are embedded in the seed of new crop varieties, structuring agricultural development. While industrialisation may boost productivity in some contexts, this thesis is looking for alternative models, and specifically aims to analyse the transformations in social relations of production and innovation that go along with the introduction of new biotechnologies.

Thirdly, as a corollary to the increasingly private nature of agricultural research, the rise of an international intellectual property system was discussed, as well as different initiatives to create 'freedom to operate', especially for the public sector in developing countries. While no direct relation is assumed between agro-biotechnological design and ownership relations, the intellectual property landscape cán have important influences on the choice of technologies that are being used, and the kind of partners that are being involved in projects of pro-poor agro-biotechnology development. This provides a third entry point to study contemporary projects of agro-biotechnology development.

Importantly, there is a certain common ground between the section on agricultural modernisation as homogeneous and externally planned affair, and the section on the organizing principles of agricultural industrialisation. Both sections raised the issue that agricultural development may be shaped by actors that themselves are external to the agricultural production itself, and lead to a homogenization in approaches. The externalization of seed production as part of appropriationism, and the transformation to a high-modernist scientific agriculture as described by James Scott refer to similar mechanisms. However, while appropriationism is motivated by an extending capitalist system of production, the belief in high-modernist scientific agriculture may have more to do with a specific scientific ideology, and bias towards simplification of agricultural production.

Contours of a positive alternative for agricultural development

The question that arises is whether a positive counter-perspective can be offered in response to the criticisms and concerns regarding agricultural modernisation and industrialisation, presented in the sections before. Some elements of such a counter-perspective have emerged, but they do not constitute a coherent picture of an agricultural development just yet. Some of the authors mentioned have made proposals for an alternative, and arguably more positive take on agricultural development, and in doing so have referred to the possibility of endogenous development (Van der Ploeg and Long 1994), tailor-made biotechnologies (Ruivenkamp 2003b, 2005), and food sovereignty (Seedling 2005; Jongerden 2008).[37] It is not always clear what the role of technological development can be in these perspectives on the future of farming, but a central theme that emerges is the ability of innovation processes to adapt to different farming styles, different localities and different trajectories of agricultural development.

In addition, a recent idea has emerged on the potential of treating farmer seed systems as 'unsupervised learning networks' (Richards *et al.* 2009). The terms supervised and unsupervised learning are adopted from the field of artificial intelligence, and roughly relate to conditions in which a network is trained to produce a specific outcome (supervised learning), or in which a network is reaching a stable state through interaction and feedback, without any *a priori* defined outcomes (unsupervised learning). The basic argument is that 'supervised learning' in agriculture has largely failed to substantially contribute to agricultural development in

[37] The term Food Sovereignty was first coined by Via Campesina and emerged in the Statement by the NGO Forum at the World Food Summit in 1996. Food Sovereignty is an approach that is largely based on the Right to Food and a right to produce. In contrast to the concept of Food Security (the right to have access to sufficient food of good quality), Food Sovereignty can be loosely interpreted as 'the right to produce your own food'. In practice, it focuses heavily on local food networks and national or regional markets (Rosset 2003; Rosset 2006; Quaye 2007).

parts of Africa, and has underutilized the experimental capacities of farmers themselves in variety management, breeding and selection on a local basis. Unsupervised learning – it is argued – might provide a heuristic for taking the innovative dynamic already present in farmer communities as starting point, which can be further strengthened and exploited by the strategic use of genetic technologies and functional genomics data. A concrete example would be the use of functional genomics data on genetic diversity in a given population of landraces that farmers use for varietal development. Genomic data may provide an insight into whether there is any genetic potential for farmer selection, or what kind of material may be introduced into the community to maximize the potential of the exchange and selection work carried out by farmers. While this approach to rethinking the premises of agricultural development requires further elaboration, and wider demonstration in practice, it does represent a valuable perspective on how agricultural development may empower local innovation processes, without prescribing a specific, homogeneous model of agricultural production or innovation.

This thesis aims to contribute to the debate on rethinking the role of technological innovation in agricultural development by focusing on the role of farmers as co-innovators. This requires technology development to be open-ended, and to leave room for experimentation and adaptation by different farmer communities, or local projects of agricultural development. In other words, the challenge this raises for contemporary projects of agricultural development, is to take local situations, capabilities, expertise and priorities as a starting point, and to harness the emancipatory potential of biotechnologies. This means that a process of creative destruction can be challenged by a careful consideration of existing production systems and seed systems. Externalization of seed breeding can be challenged by a consideration of how farmers or social movements themselves can be empowered in their varietal management and seed breeding experiments. Externally development efforts can be challenged – and complemented – by initiatives of endogenous development, that treat farmers and their communities as subjects of development, rather than as the objects of development.

This leaves us with the important question what these trends mean for technology development. In that context, it should be noted that – for example – mechanisms of appropriationism and substitutionism are not new, and neither are their associated processes of standardization of agriculture and disconnection. Goodman *et al.* (1987) trace back these processes to the early 19th century in their analysis. However, they did argue that the advent of modern biotechnology was strongly supporting these trends, and extending their scope. Industrial fermentation, the genetic engineering of micro-organisms and the use of immobilized enzyme technology were expected to further allow the interchangeability of agricultural products, and their replacement by artificial alternatives. In plant breeding, the development of varieties with specific herbicide tolerance, disease resistance, or with improved capabilities to deal with poor soils, would further concentrate the control over agricultural production in the hands of the seed breeding industry.

So the question is what the observation that biotechnologies have been instrumental in processes of appropriation and substitution means for the use of these technologies in another setting, with different goals and intended effects. If we are questioning the appropriateness of the industrialisation paradigm for international agricultural development, what does that mean for our questioning of technical design? Does this analysis of contemporary agro-biotechnology development imply that we should not only evaluate policies, strategies and approaches, but also the material design of the technologies that are being developed?

The relationship between technical design and social structures

The above has raised a discussion on the relationship between concrete technologies, production systems, and trends of agricultural modernisation. While it is generally acknowledged that there are some connections between technological design and the context in which it was developed, important differences are evident in how this relationship is conceptualized. As a result, visions of how complex it is to adapt technologies to different situations differ. In order to understand why one could have a different take on how to design 'appropriate technology', it is helpful to introduce a framework developed by Andrew Feenberg that distinguishes four different conceptualizations of technology, as summarized in Table 3.2 (Feenberg 1999, p. 9).

Feenberg starts with the common sensical notion that treats technologies as neutral means, which – within certain technical limits – can be used to reach whatever end a user of the technology may have in mind. Feenberg calls this an *instrumental vision* of technology. This is perhaps the most common-sensical notion of technology and affirms both its neutrality and its responsiveness to the intentions of a user of technology.

Next to this vision, he introduces *technological determinism* which has its roots in an optimistic believe that technical progress will ensure humanity's advance toward freedom and happiness. In this framework, technology is considered to be neutral, since it does not change the direction or 'ends' of history, but merely advances progress to an unquestioned end state. Technology itself is expected to develop according to its own inherent logic, from less advanced to more advanced stages of development.[38] This 'unilinear path of development' also implies that the means and ends of technological change are connected. Since technological progress remains unquestioned, it is society that has to adapt to the rules and effects of technological development (Feenberg 1999, p. 77-78).

The optimism of technological determinism is countered by a tradition of romantic protest against technologies and mechanisation. Feenberg refers to *substantivism* which holds that

[38] Feenberg points at the influences of Marx and Darwin that can be seen in technological determinism. Both had a theory in which biological evolution or modernizing society had it is own logic of progressing from primitive to more advanced stages, which is echoed in technological determinism.

Table 3.2. Four visions on technology (Adapted from: Feenberg, A., (1999), page 9). The numbers printed in the background refer to the order in which the four visions are discussed in the text below.

	Autonomous	Humanly controlled
Neutral *(complete separation of means and ends)*	Technological determinism *(unilinear path of progress according to inherent logic)* 2	Instrumentalism *(technology is what it becomes in the hand of a user)* 1
Value-laden *(means form a way of life that includes ends)*	Substantivism *(means and ends linked in systems, inherent inclination to power and control)* 3	Critical theory *(choice of alternative means-ends systems)* 4

technologies not only have an instrumental, but also a substantive content, loaded with social meaning. In this view, technology is not neutral, but embodies certain values. Like in determinism, means and ends are connected: *"how we do things determines who and what we are"* (Feenberg 1999, p. 2). The main difference is its much more pessimistic view of the ends of technological change, since it assumed that technology was in the hands of a powerful elite within society, protecting their interests, and is therefore inherently biased towards domination. The only way out for substantivist thought was the rejection of technology and a return to nature or arts.

However, Feenberg argues that from within this pessimistic tradition of criticizing technological development also scholars emerged that perceived technologies as forms of power and control, but argued that technical domination is related to social organisation.[39] In other words, it is related to man-made social configurations and could therefore be 'reconstructed' to play a different role in different social systems. While this sounds a bit similar to an instrumental view of technology, the difference is that choices are not at the level of particular means, but at the level of whole means-end systems. In contrast to substantivism and technological determinism, such a theory of technology steps away from the essentialism in these frameworks, claiming that there is not just one 'essence' of technology. This way, they opened the door to a *Critical Theory* that is characterised by a more constructive criticism and democratization of technological development.

[39] See e.g. Marcuse (1964) and Foucault (1977).

How technological design relates to context of application – some examples

Critical theory in its constructive form provides opportunities for emancipatory change of technological means-end systems. However, an important condition for this emancipatory action of technological modification is that its entanglement with social relations of power is made visible in the first place; something that is not always quite self evident. According to Feenberg, this blindness to the entanglement of technology with relations of power is primarily due to a 'hegemony of technical rationality', which means that technologies are strongly perceived in terms of function and cost-effectiveness and much less in terms of what their social meaning is, or how they structure our society in different ways (Feenberg 1999, pp. 86-87). We commonly perceive a distinction between technologies which have a technical function, and their social meaning which depends on the ideas and intentions of a user of the technology. Feenberg argues that this separation between 'technical function based upon technical design' and 'social meaning based upon its use' is not tenable, and masks the fact that technologies in their very material design already embed specific social meanings.

To borrow one of his examples: the size of machines in most European factories today prohibits their operation by children's hands; they're simply too big. This is considered to be self-evident, since we have abandoned child-labour in Europe quite some time ago. However, halfway the 19th century, this was not self evident at all; children were commonly working in factories and the machines were, by their very design, adapted to this social condition. This illustrates that the technical design of these factory machines not only has a technical function (can it efficiently be operated?), but also a social meaning (who is to operate these machines?) (Feenberg 1999, p. 86-87). Perhaps without realizing it, technology designers have important and far reaching social power, by inscribing technical objects with social meaning, which is most appropriate in their eyes.

This relationship between technical design and a specific context of application is also apparent in the context of food production. The discussion of the organizing principles of agricultural industrialisation – appropriationism and substitutionism – already provided some proof that technologies not only have a technical function, but can also have a structuring effect on the socio-economic context of application. And, as already mentioned, hybrid seeds not only increased yields, they also fundamentally changed the agricultural production chain, and firmly embedded public or private external seed suppliers in the line of production (Kloppenburg 1988). Similarly, modified glucose isomerase enzymes did not only allow for a conversion of glucose into fructose, they created an interchangeability of sugar cane for corn syrup, profoundly restructuring the international production and trade of sweeteners (Ruivenkamp 1986; Hobbelink 1991).

These examples in fact reflect a struggle for power in the agricultural production system. To put it bluntly, hybrid seeds shift the power balance towards seed suppliers who become essential partners in production. Enzyme technology shifted power to food manufacturing industry that benefits from the interchangeability of its inputs for production. But technological development can also more subtly be related to a specific idea of how to organise agricultural production. Here, technologies may not have such a direct effect on the power balance in agricultural production systems; instead they may implicitly support one or another view of agricultural development. This effect is best demonstrated by the example of breeding for disease resistance. Both conventional farming systems, as well as organic farming systems have to deal with insects, viruses and fungal diseases. Hence, generally speaking a trait as disease resistance is of general use and applicability. But a closer look at the understanding of disease resistance reveals important differences in conventional and organic approaches, and different criteria for variety development.

Different types of disease resistance in crops exist. In general a distinction is made between (1) monogenic complete (or vertical) resistance, (2) polygenic (often partial or horizontal) resistance, and (3) different resistance genes in a multiline variety or in a variety mixture (Louwaars 1997). While complete resistance is a very useful trait, its function generally depends on a single gene and is therefore relatively vulnerable. New viruses that evolve through mutations may develop a way around the plant's resistance mechanism. This means that new crop varieties with a complete resistance trait are very interesting because of their excellent disease resistance, but they are only expected to last for a limited number of seasons. Seed breeding companies are constantly working on new resistance traits to anticipate the breakdown of resistance traits in currently marketed varieties. The 'need for speed' this generates in the plant breeding industry is met by increasingly powerful plant breeding strategies and technologies, like transgenics and marker-assisted breeding.

For the organic sector, disease resistance is important as well, but it is generally approached in a different way. The agro-ecology of crop production as a whole is considered, and the objective is to reach a balance in the crop-environment interaction (Lammerts van Bueren *et al.* 2002; Lammerts van Bueren 2006). By consequence, the organic sector is explicitly looking for sustainable resistance traits, and general 'robustness' of the crop. This means that for long term breeding programmes, resistance traits based upon the introgression of a single resistance gene are not preferred. Rather, more complex, quantitative resistance mechanisms are favoured that may not result in full 100% resistance (less absolute resistance), but do provide the plant with a more durable resistance against diseases or pests (Lammerts van Bueren 2006). Such traits are difficult to breed using genetic modification since they generally involve a wide range of genetic factors, rather than one or a few sharply defined genes. Molecular markers may in fact be useful in this context since they can help combining genetic elements that are partly contributing to a desired trait. However, the different crop ideotype for organic cultivation (Lammerts van Bueren *et al.* 2002) would require the development of markers for entirely new traits. Moreover, in organic farming conditions, where one is less capable of controlling the environment than in conventional farming systems, the link between DNA (genotype) and actual traits of the plant (phenotype) is expected to be less consistent. Instead, the crop-environment interaction is expected to play a much larger role in organic agriculture than in conventional agriculture, dramatically decreasing the predictive value of the presence of certain DNA elements for the successful expression of a certain trait or quality in the field. This implies that plant breeding for the organic sector does not only focus on a different crop ideotype with different traits, also the requirements for technologies in organic plant breeding are different. While molecular markers may prove to be of some interest in organic plant breeding, there is relatively strong need for traditional field testing in which field trials mimic the conditions in organic farming practice, even though this strategy does require a longer period of testing and selecting.[40]

These differences in how disease and resistance to disease are conceptualized in different types of agricultural production, lead to entirely different visions of what kind of technology is appropriate for agricultural development. This supports the earlier suggestion that technologies for agricultural development can only be discussed and evaluated in terms of their technical functioning within a given production system.

The politics of technological design

The discussion above has made clear that there is an important relationship between technical design, context of application, and perspective on agricultural development. But this raises an important question that has so far remained unanswered: does a certain technology by definition have a prescriptive function in a given production environment? If we criticize the role of biotechnologies in an industrializing agricultural system, does that mean that these biotechnologies cannot have a different social meaning in another context? Does a change in social meaning require a concrete redesign of the technology, or can technological design gain different meanings in different settings? In other words: to what extent does the technological artefact itself have a structuring function, across different contexts of application?

These questions hit a fundamental discussion in Science and Technology Studies (STS) and the Philosophy of Technology (see e.g. Woolgar 1991). On the one hand, these traditions stress the way in which technical design reflects social norms, and that technological design

[40] This paragraph focuses on rather technical arguments of why organic plant breeding has a different take on breeding for resistance, leading to a different crop ideotype, different traits, and different plant breeding technologies. In addition, the organic sector takes the concept of 'naturalness' as starting point in breeding and cultivation, and attaches importance to the intrinsic value of crops, based on their autonomy, wholeness or completeness, their species-specific characteristics and their being in balance with their species-specific environment. These criteria strongly object the use of genetic modification in organic plant breeding, and the use of other invasive genetic technologies such as protoplast fusion. See Verhoog *et al.* (2003) and Lammerts van Bueren and Struik (2004) for a discussion of the concept of 'naturalness' in organic agriculture, and the consequences for which breeding technologies are acceptable and not.

and application is thoroughly contingent upon social, political and institutional context (e.g. Bijker 1995; MacKenzie and Wajcman 1999). This notion has been powerfully captured in the term 'socio-technical ensemble' which expresses the complete entanglement of the social and technical (Bijker 1995, p. 274). But a number of scholars has also very explicitly drawn attention to the coercive power of technological artefacts, and the notion of 'the politics of technologies.' Langdon Winner addressed this issue head on in his seminal article "Do artifact have politics?" (Winner [1980] 1985). He argued that technological design can enforce social norms by allowing or denying certain social groups access to specific technologies or services. Moreover, he argued that technology can be highly complementary with certain types of social organisation (e.g. hierarchical or centralized management).

A few years later, Madeleine Akrich and Bruno Latour have discussed the delegation of morality to artefacts in terms of a 'script of technologies' which allows apparently mundane artefacts (door closer, seat belt) to implicitly enforce certain social norms (Akrich 1992; Latour 1992). Also Andrew Feenberg builds upon this tradition when he introduces a notion of a 'technical code' which brings technological artefacts in accordance with the social meaning they have acquired. Technologies are argued to materialize social norms and ideologies, which become embedded in the material design of the technology. This implies that technologies can also be prescriptive in the kind of social relations they mediate. As far as Feenberg is concerned, this has an important implication for the democratization of technological design. Not only should technological institutions be governed in a democratic way, *also the material design of technological artefacts should be opened to reconstruction* in the course of a democratic involvement of citizens in the socio-technical shaping of our world, ultimately leading to what Feenberg calls a "Deep Democratization" (Feenberg 1999, pp. 142-147).

But there are two problems with this understanding of the politics of technological artefacts. One is conceptual, the other pragmatic. First, there is a potentially problematic interpretation of the concept of a 'technical code' of technologies, since it may seem to suggest that a technical object has a specific technical configuration, which leads to a specific social effect across different social and economic contexts in which the technology may function. Winner already countered that interpretation by mentioning that *"A ship out at sea may well require a single captain and obedient crew. But a ship out of service, parked at the dock, needs only a caretaker"* (Winner [1980] 1985, p. 37). In other words, context of application matters.

Bruno Latour similarly stresses that the architecture of buildings from the Belle Epoque successfully separated the servants from the bourgeois in the house. But the same building today has the *"perverse tendency to force the students inhabiting its coveted 'chambre de bonnes' to climb six stories through a steep and narrow staircase, while the happy owners of the flats are allowed to glide through a comfortable lift inserted inside a wide staircase."* (Latour 2004). This of course is a totally unintended but discriminatory effect of the architecture from an entirely different epoch. Historical context matters too.

In addition, Brian Pfaffenberger makes a powerful and useful argument on the importance of culture in the way technologies gain a specific coercive force. To him, technological artefacts have affordances which are inherently multiple, depending on the perception of users or affected stakeholders. Pfaffenberger argues that it is discourse ('ritual') that privileges and legitimizes a specific interpretation of technologies, thereby constituting a political effect. *"The artifact embodies political intentions, but these intentions do not come to life in the absence of ritual"* (Pfaffenberger 1992, p. 294). He illustrates his point with reference to the plain Victorian hallway bench on which servants had to await the master of the house, in the nineteenth century. A myth of hygiene mystified and legitimated the use of a plain and hard bench, which was supposedly not to humiliate the servant class, but because they would only soil nicer benches with dirt from the streets. Today, many antique collectors place Victorian hallway benches in their homes, but with a very different intent. According to Pfaffenberger: *"What made the hallway bench into a political artifact in the nineteenth century was the ritualization of the hallway space: Profound decorum standards called for members of the masters' class to be admitted straightaway into the interior of the house, while members of the servant's class were seated on the bench, signifying their inferiority"* (Pfaffenberger 1992, p. 294). The same benches today have an entirely different social meaning, expressing a certain taste in interior decoration, rather than the inferiority of guests that are to be seated on them. In other words, not only historical context matters, ritual matters too.

The conceptual conclusion must be that technologies are profoundly political, but that their political meaning depends upon the social and historical context they are part of, and the rituals and discourses they are surrounded by.[41] This takes us away from an essentialist understanding of the political nature of technologies, in the sense that we do not any longer see technologies as having only one particular political function or 'meaning', captured in their technical code. So, rather than considering a technology in isolation, its socio-political meaning should be studied within a specific context of application. The discussion above has made clear that such an analysis should both involve the social dynamics in that context, as well as the material design of the technology itself.

As announced, a second – more pragmatic – problem with our understanding of the 'politics of technologies' emerges in the context of this research. The conceptual understanding of the

[41] There is a more fundamental critique of Winner and the notion of the 'politics of artefacts' which deals with the asymmetry in considering some technologies socially constructed, while at the same time treating other technologies as having a specific meaning, code or script. See e.g. Woolgar (1991), Grint and Woolgar (1995), Joerges (1999), and Woolgar and Cooper (1999). This discussion has some profound methodological implications for science and technology studies, and calls for an increased reflexivity in which the relativism of the construction of technology is also applied to the analyst herself/himself. This discussion is considered to be beyond the scope of this chapter and thesis; however the contextualized understanding of 'the politics of technologies' as here proposed is thought to meet the expressed concerns to an important – workable – extent.

coercive and prescriptive force of technology is directly related to a very concrete, material object with a specific technical design. In discussing agricultural modernisation and the role of modern biotechnologies, we do not always discuss a specific technical object, but rather a research trend, or breeding strategy. This first means that we explicitly have to consider what the material dimensions are to contemporary biotechnology or genomics development. But it also means that we have to look further than concrete technical objects, and need to consider, for example, the 'politics of breeding strategies'.

Paul Edwards provides an interesting and helpful insight by elaborating the relationship between *infrastructures* and modernity (Edwards 2003). He extends Langdon Winners' argument on the politics of technology to include wider sociotechnical systems, or infrastructures: *"infrastructures act like laws. They create both opportunities and limits; they promote some interests at the expense of others"* (Edwards 2003, p. 191). [42] It makes perfectly good sense to treat an agricultural production system, and notably a seed supply system as an infrastructure, consisting of different components and institutions allowing it to work, and to provide farmers with agricultural inputs. This understanding allows us to reflect upon the political nature not only of technical artefacts, but of the infrastructure –the sociotechnical system – they are part of in a wider sense.

In conclusion, in order to understand the significance and 'social meaning' of technological design, the challenge is to study the role of technologies as part of an (agricultural) infrastructure, rather than as isolated technical artefacts. This agricultural infrastructure provides the context of application, the historical context, and the ideology of agricultural development, needed to understand and evaluate the role of biotechnologies within this infrastructure. This contextualized understanding of the social meaning of biotechnologies in international agricultural development will be central in the description of case studies in the upcoming chapters.

In conclusion: key elements of a conceptual framework

The discussion before on the wider historical trends in agricultural modernisation and industrialisation, and the instrumental role of (bio)technologies in those processes, has provided us with an additional dimension of questioning the trajectories of agricultural development. The consecutive discussion on the relation between technical design and its meaning in the wider context of application has made clear that questioning trajectories of agricultural development also requires a questioning of agricultural technologies themselves.

[42] Edwards stresses that while infrastructures are commonly perceived in terms of 'hardware', we should acknowledge that they are in fact *socio*technical in nature. He writes: "Not only hardware but organisations, socially communicated background knowledge, general acceptance and reliance, and near-ubiquitous accessibility are required for a system to be an infrastructure" (Edwards 2003, p. 188).

This brings us back to the discussion, started in the introductory chapter, on what it is that makes biotechnologies 'appropriate' for international agricultural development.

In relation to this discussion on 'appropriateness', a concern was expressed that the adaptation of biotechnologies for agricultural development is not as self-evident as it may appear to be, and that a superficial adaptation of biotechnologies might mask underlying processes of social transformation, that remain implicit in agricultural and technological development. The discussion in this chapter has provided various elements of conceptual framework which allows a further explication of this concern. This final concluding section will summarize and refine these key elements.

Reflexive development and reflexive technology design

The notion of 'reflexive development' was introduced in the first chapter to indicate a process of reflecting upon and responding to the effects of development efforts and the comments and criticisms it invokes (Nederveen Pieterse 1998). As mentioned before, in contrast with polarized visions of mainstream-, alternative, and post-development, this notion highlights flexibility and a learning dynamic which arguably more accurately describes how development efforts have changed over the years and have – for example – been influenced by the work of civil society organisations. In practical terms, the term reflexivity describes the degree to which project managers or scientists reflect upon their role in an innovation process and adapt the projects goals and methods in order to optimise its potential impact.

This focus on reflexivity in the context of agro-technological development relate this research to ongoing debates on the potential for reflexive technological development, or 'reflexive design' (see e.g. Schot 2003; Grin *et al.* 2004; Bos 2008). Here it is not so much a specific area of agro-technological development that is of interest, but rather a methodology for engaging with technological development. According to Bos, reflexive design is *"a specific form of deliberative or participatory technology assessment oriented towards the definition of both the problem and the solution in a reciprocal argumentative exchange between the actors involved in the problem"* (Bos 2008, p. 36). This notion of reflexive design relates to a wider shared interest in the democratization of technological development, and the potential for the wider involvement of various stakeholders and different types of expertise (e.g. based upon scientific knowledge or based upon experience). The definition of reflexive design as formulated by Bos is interesting because of its emphasis on the deliberative definition of both the solution and the problem at hand. Moreover, Bos stresses that in this deliberative process, more is needed than negotiation and trade-off between different interests. In addition *"institutionally and technologically embedded assumptions, norms, knowledge claims, distinctions, roles and identities that are normally taken for granted must now be critically scrutinized"* (*ibid.*).

These descriptions of what reflexive design implies provide some starting points for an operationalization of the concept in the case study analyses. It suggests a questioning of the degree to which various actors in the cases reflect upon their own institutional context, the interests at stake, and the assumptions underlying their work. Having said that, it would be very hard to provide a quantitative measure of the degree of reflexivity of the relevant actors in a given project. Only a rough indication may be obtained from conducting interviews with these actors. However, rather than measuring the quantitative degree of reflexivity, the main interest of this research is in finding out to what kind of considerations it leads. A critical reflection upon the project may – for example – involve a reconsideration of its technical outputs or the degree to which it is needs based. However, the study is specifically interested in learning whether reflection is also being made regarding the agricultural production system that is implicitly or explicitly supported by the project, and what that means for the social roles and responsibilities of farmers, private sector input suppliers and (other) technology developers. In other words, rather than a quantitative measure, the notion of reflexivity provides an entry point for questioning the dimensions in which a project is shaped, reconsidered and optimised in order to provide a contribution to 'appropriate development' for 'resource poor farmers'.

Pro-poor and Appropriateness

This study specifically focuses on projects that have a pro-poor focus and that aim to contribute to poverty alleviation. This concept of a pro-poor orientation is also guiding for the notion of 'appropriateness', since it defines *for whom* the project or technology should be appropriate. In general, this term 'pro-poor' is commonly used to express the ambition of a project to address rural poverty. However, in spite of its relevance and common usage, the term is clearly vague and ambiguous. Paradoxically, this makes the use of the term pro-poor both problematic, as well as illustrative for a discourse in which 'the poor' are frequently treated as an abstract and decontextualized category. One way of dealing with the use of this term in this thesis would be to postulate a more robust definition, elaborating who exactly are the supposed beneficiaries of the genetic technologies the thesis is studying. However, for this study it is expected to be more fruitful to treat the category 'pro-poor' as something that exists because people use it to define their own work or project. Rather than delineating the category of 'resource poor farmer' (as beneficiaries of the studied projects) in terms of income or standard of living, the category has a meaningful existence as an entity that is constructed differently in different projects and different contexts. It is precisely these differences in identifying what makes a project 'pro-poor', or how it is going to reach 'resource poor farmers', and who these people are, that can provide valuable insights for the study into the different ways in which projects are reflexive in their attempts to contribute something 'appropriate'. Hence, in the rest of this thesis, the term 'pro-poor' will be consciously used as a term which is ambiguous in representing a clear target group of people, but which is rather precise in indicating the general orientation of technological projects, that aim to contribute something to agricultural development in developing countries. More than its general definition, what matters is its operationalization in concrete case studies.

A similar issue arises with the notion of 'appropriateness' when studying technologies that are being developed or projects that are being set up. As has been discussed by the end of Chapter 1, it is not feasible nor fruitful to postulate a single conclusive definition of what makes a technology or a development process 'appropriate' for a specific group of beneficiaries (resource poor farmers in our case). Rather, the term is used as an empirical lens or entry point to start questioning how in different projects this notion of 'appropriateness' is operationalized in practice. Rather than in a clear definition, the interest for this study lies in the expected diversity of the concept.

Politics of technology and Reconstruction

Apart from discussing a wider context in which agricultural development and modernisation takes place, this chapter has introduced the notion of 'politics of technologies'. This term implies that, next to a technical function, biotechnologies have a social meaning within specific production systems, which allows them to restructure social relations of production and innovation in agriculture. As discussed in the previous section, it is important to study the role of technologies as part of an (agricultural) infrastructure, rather than as isolated technical artefacts, in order to understand their role in social structures. This agricultural infrastructure provides the analyst with the relevant context of application, a historical context, and the underlying ideology of agricultural development, allowing the researcher to understand and evaluate the role of biotechnologies within this infrastructure. In studying specific cases of technology development in the case studies in Chapters 4 to 6, reference will be made to this notion of the politics of technical design in order to discuss how the specific material design of technologies is adapted to suit the objectives and context of the project, or – on the contrary – how this design implicitly introduces or changes social structures that remain unquestioned.

Importantly, the exploration of the relationship between technical design and social meaning has opened the door to a more profound questioning of technological artefacts themselves. This kind of analysis requires the introduction of new language in order to distinguish between different levels of making technology 'appropriate'. It is now clear that adapting technologies to perform a specific role in agricultural development can relate to the technical functioning of new technologies, but that it can also involve questioning and challenging the social relations of innovation, and the roles of scientists, breeders and farmers. In order to indicate such instances in which adaptation and reconfiguration of a project goes beyond instrumental technical adaptations, and includes a more radical and profound questioning of socio-technical configurations, the term 'reconstruction' will be used in this thesis.

The term reconstruction is adopted from Andrew Feenberg (1999) who uses the term to indicate a process of 'reconstruction' of social configurations in which both the technical means, as well as the societal ends are reconsidered. His definition highlights a contrast between viewing technology as a neutral tool (leading to instrumental adaptations), or viewing technology as essentially interwoven with social systems (requiring a more profound rethinking of how both technical design and social systems can be changed). The use of the term reconstruction implies that not only the technical function of a technology changes, but also the roles and responsibilities of scientists, breeders and farmers working with the technology may change. This becomes especially relevant against a context of different – and rivalling – agricultural production strategies as described by Pistorius and Van Wijk (1999), which represent entirely different pictures of how agricultural production is supposed to be organised, and what different roles of the state and private sector are in that development. Technological development within these different strategies for agricultural production does not only require a questioning in terms of their technical functioning, but also in terms of their relation to a specific development strategy. It is a market-led industrialisation strategy in which genetic breeding technologies have become a crucial tool for innovativeness, but the question is whether they can also have an added value within a non-industrializing and farmer-centred strategy for agricultural production, and what kind of technological adaptations and reconstructions that would require.

Externalization in production- and innovation systems

The notion of externalization has been extensively discussed in this chapter against a background of agricultural industrialisation. It refers to the degree to which – in a specific project – farmer autonomy is conserved, strengthened or reduced in terms of direct accessibility to their means of production (i.e. their agricultural inputs). Externalization implies a loss of such farmer autonomy. As a corollary to externalization, homogenization of farming styles is described as a common consequence. Both concepts are the outcomes of a mechanism of appropriationism (Goodman *et al.* 1987; Ruivenkamp 1989, 2003a), as well as the result of a high modernist approach to agricultural development, as described by Scott (1998), Jongerden (2008) and Van der Ploeg ([1999] 2003, 2008). The process of externalization is an example that rather clearly emerges out of this kind literature as an implicit organizing principle of agricultural industrialisation and modernisation. What makes it interesting for the questions asked in this study is that it generally remains implicit and unquestioned, while the authors here referred to all stress the profound consequences for farmer production systems in terms of dependency relationships and homogenization of farming styles.

Interestingly, in the literature reviewed in this chapter the notion of externalization has generally been discussed with reference to the social relations in a *production system*. However, a similar mechanism can be identified in the context of *innovation systems*, in which the innovative capacity is being externalized from the realm of farmers to external technology developers.

Rather than questioning the direct access to the means of production, externalization of the innovative capacity relates to the degree to which downstream research partners, or end users can play a more – or less – significant role in determining the exact technical functioning and social meaning of the technology. In this context a dichotomy between treating farmers as 'recipients of technology' or as 'co-innovators' has already been introduced as a heuristic for the different social roles in an innovation process. Both terms refer to different approaches to engaging with agricultural and technical development, and with the conceptualization of farmers (or other local stakeholders) in that process.

Concluding remarks

These elements constitute the conceptual heart of the case study analysis in the upcoming chapters. They provide a theoretically informed entry point to data collection and analysis. At the same time, they are nothing more than sensitizing concepts that allow for a first approach of the three cases (Bowen 2006). From the cases new concepts will emerge that provide new insights into how reflexivity works in making different projects appropriate for resource poor farmers in different ways. In response to the criticisms of a homogeneous, and externally imposed model of agricultural modernisation, the focus will be on the emancipatory potential of agro-biotechnologies, and the ways in which they can link up with, or hybridize with existing production systems and informal seed systems.

Chapter 4

Transgenic insect resistance for the poor – Towards a win-win-win situation in Indian vegetable farming

> *"I like to call it win-win-win. The company wins – of course –, the farmer most definitely wins as well, and so do the consumer and society at large, because we'll get rid of the poisonous residues in the vegetables. If we'll handle this project the right way, there will only be winners"*
>
> (anonymous manager involved in CIMBAA consortium, November 2006)

Introduction [43]

This chapter elaborates the case study of a project aimed at the development of a transgenic cabbage variety in India, in which both public and private partners aim to make the technology appropriate for vegetable production by small scale, resource poor farmers. The operationalization of 'appropriateness' in this project is interesting and complex, since the project has multiple objectives, and some of those remain rather implicit. Next to delivering a transgenic crop with effective insect resistance, the project hopes to address commonly expressed public concerns about transgenic technology, and thereby to avoid the controversy that many projects working on transgenics have met in the past. This second objective requires the project to be sensitive to potentially controversial aspects of its outputs, and to adopt a reflexive approach in which the project anticipates and responds to critical concerns. This approach is reflected in a variety of ways in which the technology and the strategy for commercialization are reconsidered and modified during the course of the project. However, in spite of this reflexivity, the analysis of the project also reveals important underlying stakes in supporting the externalization of seed supply to specialized companies and institutes. The discussion of the case study in this chapter will focus on the various dimensions of technical and organisational adaptation, but will also question how the project reflects and reinforces certain assumptions regarding the roles of seed suppliers and farmers in agricultural innovation.

---

[43] This chapter is largely based upon material presented at the Development Studies Association Annual Conference 2007, University of Sussex, UK, 18-20 September 2007. The accompanying paper has been published as: Vroom W. (2008). 'Redesigning biotechnology: experiences of a public private partnership in the development of pro-poor transgenic cabbages in India.' *The European Journal of Development Research* **20**(3, September 2008): 398-414.

Concretely, the case study addresses (1) how the project aims to make the technology and the mode of commercialization appropriate for small scale vegetable farmers, (2) the extent to which the project takes the perspectives of farmers and other stakeholders on board, and (3) the structuring influence of intellectual property and biosafety regulations on the setup of the consortium, and the repercussions for the central position of the patent owner. Questioning the project on these levels provides insight in how the project not only is committed to provide a technically sound solution to the pest infestation in Indian cabbage production, but how it also structures the future innovation and production system in a specific way.

Finally, it is important to note that this chapter is not evaluating the project in terms of successfulness of the final technological solution as such, nor making a wider argument pro or against the use of transgenic crops. Instead, what is elaborated is the extent to which technological development is responsive to different needs and circumstances, and to what kind of adaptation processes that may lead.

The setup of a public private partnership for the development of Bt Brassica

Around the year 2000, several parties in India were looking for a sustainable solution to pest problems in *Brassica oleracea* (leafy cabbages, including white cabbage and cauliflower), primarily caused by infestation by the diamondback moth (*Plutella xylostella*). India is one of the major producers of cabbage and cauliflower worldwide with approximately 0.483 million hectares under production, producing about 6.335 million tons per annum (Mohan and Gujar 2003). The diamondback moth causes an annual loss estimated at about $16 million on the basis of 25% damage, and forces farmers to apply frequent sprays of insecticides, adding up to 38% of cultivation cost in cabbage and cauliflower (Shetty 2004). Since no resistance against this insect in wild relatives within the Brassica family is known, the only way of making the crop itself more resistant to the insect attacks, would be to introduce transgenic insect resistance.[44] The British National Resources Institute (NRI) had previous experience with transgenic Bt cotton[45] in the country, and was investigating the opportunities of setting up

[44] Obviously, next to making the crop itself more resistant to the moth, other strategies could have been followed, but host resistance was quickly defined as an interesting and efficient avenue to follow. This choice will be discussed later in this chapter.

[45] 'Bt crops' are transgenic crops carrying and expressing genes that originally occur in *Bacillus thuringiensis*, a soil bacterium. The presence of Bt genes leads to the expression of an endotoxin that specifically kills certain groups of insects, without being harmful to other non-target organisms (Schnepf *et al.* 1998; Shelton *et al.* 2002).

Integrated Pest Management (IPM) systems for Brassica in which transgenic Brassica with Bt resistance would be an important element.[46]

Nunhems India Pvt. Ltd., a major Indian seed company, was already working on Bt Brassica in India at the time. Nunhems India is a part of the Dutch Nunhems Seeds corporation, which in turn is a subsidiary of Bayer CropScience. At the time, Nunhems India was working with a specific Bt gene: the Cry9C gene. However, this Bt gene had been involved in a controversy over the contamination of corn intended for human consumption with transgenic corn that was not intended for consumption, produced by the US company Starlink (Bucchini and Goldman 2002). The public controversy over the use of this gene made the seed company decide to abort the programme of Bt Brassica development, in combination with the recognition that getting any concrete products through the regulatory process would be very difficult and expensive, and the fact that the technology used was not up to the latest technological standards anyway. All in all, the development of transgenic vegetables in India, though technically feasible and interesting, was at the time considered to be commercially unattractive.

However, several people involved at the time, both at the seed company, as well as at NRI believed that the development of Bt Brassica could significantly reduce the problems with diamondback moth infestation that many Indian farmers were facing and which led to exorbitant high pesticide usage. Next to a commercial goal, the development of insect resistant cabbages was thought to have an important potential for agricultural development, in terms of lowering production costs for farmers, reducing environmental pollution because of pesticide usage, and reducing both farmers and consumers poisoning because of frequent pesticide applications and high levels of pesticide residues in fresh vegetables. These potential benefits were deemed worthy of public support, which led to the setup of a public private partnership for the development of Bt Brassica for Indian farmers.

The setup of this project was not without challenges. One of the initiators of the project, Bert Uijtewaal from Bayer CropScience has reported how a number of potential sponsors initially expressed their interest in the initiative, *"but were nervous about joining a project involving the development and release of transgenic material"* (Uijtewaal 2006, p. 218).[47] Another drawback mentioned by Uijtewaal was the expected duration of the project, since it was expected to take

[46] For a review of historical perspectives and contemporary development in IPM, see Kogan (1998). Kogan also provides the following definition of IPM: *"IPM is a decision support system for the selection and use of pest control tactics, singly or harmoniously coordinated into a management strategy, based on cost/benefit analyses that take into account the interests of and impacts on producers, society, and the environment."* (Kogan 1998, p. 249). In practice this primarily means that biological and chemical means of insect control are combined, instead of fully relying on chemical pest control.

[47] These included European development agencies, but also the Food and Agriculture Organisation (FAO) of the United Nations and the Agricultural Biotechnology Support Programme (ABSP II), which is funded by USAID (interview data).

at least 8 years from the start of technical work, to the release of developed plant material. Nonetheless, in 2002 the negotiations with various potential partners and donors led to the establishment of the 'Collaboration on Insect Management for Brassicas in Asia and Africa' (CIMBAA).[48] The consortium is formally headed by AVRDC (The World Vegetable Centre, Taiwan), and includes the Centre for Environmental Stress and Adaptation Research (CESAR) of the University of Melbourne, Cornell University, and the National Resources Institute (NRI) of the University of Greenwich (UK) as public partners. In addition a number of research partners are involved that address specific research aspects of the project. Nunhems India Pvt. Ltd. is the only private partner in the consortium (see Figure 4.1 for an overview of the consortium structure). Financially, the project is supported by Nunhems (which pays for the development of the gene constructs, the transformation, field trials and part of the safety analyses), and by public sector money: the project is divided in smaller subprojects for which funding of various governments or donors is attracted.

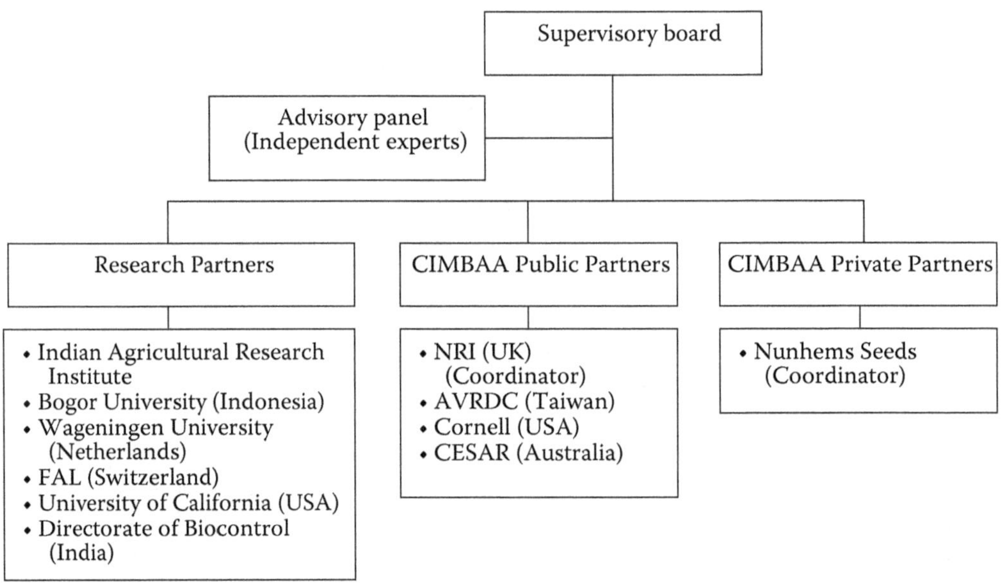

Figure 4.1. Structure of the CIMBAA – consortium (source: www.cimbaa.org; last accessed 17 September 2008).

[48] See http://www.cimbaa.org (last accessed 17 September 2008).

Phases of development

The project is organised in several phases of development. The first phase of the project focused on formulating a theoretical technical solution, with a panel of internationally renowned scientific experts. This solution would have to address several issues regarding the safety of the product, but most notably also regarding the build up of resistance of the diamondback moth (DBM) to the Bt endotoxins produced by the transgenic crops. The insect has a history of quickly developing resistance to a range of chemical pesticides, and is also expected to develop resistance against Bt foliar sprays (Talekar and Shelton 1993). These technical challenges led to a design of a transgenic plant with a dual Bt gene construct, expressing two different Bt endotoxins. Lab tests had shown that resistance build up by the insect to two toxins at the same time would occur much slower, than to a single toxin (Zhao *et al.* 2003). Moreover, by constructing the two Bt genes in a tandem construct, they can only be transferred together in any future crosses, preventing any reduction in protection because of the potential loss of one of the two Bt genes. Credibility for this technical design was build by discussing the robustness of the design with a wide range of scientific experts, and asking them to express their support for the design by signing a public scientific statement.[49] Importantly, the proposed technical solution included that the Bt seeds were to be used in an IPM context, in order to prevent the build up of resistance over a longer time frame, and to address any problems with secondary pests. Although DBM is the most devastating pest for Brassica at this moment, secondary pests also play a role. Moreover, experiences with Bt cotton in China seem to demonstrate that the elimination of the primary pest can lead to the rise of secondary pests that become more problematic (Wang *et al.* 2006). For this reason it is considered to be important to have an IPM strategy in place which not only focuses on the primary pest, but on a wider range of plant-insect relations.

The technical design with a dual Bt construct was operationalized in a following phase, focusing on the building of a genetic construct, the transformation of plants and selection of 'elite transformation events' in which the incorporation of the gene construct had led to a stable insect-resistant plant that shows no side effect that may be caused by the genetic modification. By the end of 2006, a first series of contained field tests with these genetically transformed plants had been finished, and successful host plant resistance has been demonstrated.[50] In a parallel track, two studies to the socio-economic benefits of Bt Brassica for Indian and Indonesian farmers have been conducted in 2004[51], and projects with public partners are

[49] The list with signatories of this statement is available on the project's website: http://www.cimbaa.org/support_scientists.html (last accessed 17 September 2008). Moreover, the technical concepts of CIMBAA have been presented and discussed at several entomological conferences.

[50] Presentation of Joachim Schneider, head of Bayer Bioscience, at the Biovision conference, March 13th 2007, Lyon.

[51] These reports are available at the CIMBAA website: http://www.cimbaa.org/index-2.html (last accessed 17 September 2008).

initiated studying the potential crossing of transgenic pollen with wild relatives, and the models for dealing with the transfer of intellectual property to project partners or future sub-licensees. The extensive biosafety testing that is required for commercialization of transgenic crops in India and elsewhere will be conducted once the final elite lines are developed.[52] Studies on the potential integration of the transgenic material in IPM practices are being started now the first field trials have given a proof of concept. Although the extensive experimentation with an IPM module is only possible once varieties have been developed and released for large scale field trials, the consortium has planned the assessment of potential technologies and practices that can constitute an effective IPM strategy in parallel to the variety development. Field trials with farmers and education efforts to make farmers familiar with the IPM strategy are part of the last phases of the project. While this implies that a lot of work is done in parallel, the choice of most appropriate Bt genes and the development of a genetic dual construct was all conducted in a first phase, separate from the development of an IPM context. Lessons and experiences from the application of Bt Brassica in an IPM context may therefore only be taken on board in a future generation of new transgenic material.

After commercialization, the project will enter a new phase in which the focus is on 'stewardship'. The sustainable release of transgenic Brassica requires ongoing research into the build up of insect resistance against the Bt toxins, and potential gene flow through crossing with wild relatives. Moreover, additional funds may be required for training of farmers or to provide micro-credits which can allow farmers to invest in higher quality seeds. For these purposes, plans are being made to set up a 'stewardship fund' that will take care of these issues after commercialization. The money that needs to go into this fund will have to be contributed by the sub-licensees of the technology, although a final model for this part of the project is still under development at the time of writing.

The material reconstruction of transgenic technology

The use of transgenic technology in the CIMBAA consortium is influenced by the specific objectives of the project. Not only does the project have objectives in terms of delivering an efficient means of beating insect infestation, but the project also has objectives in terms of legitimizing its existence, the involvement of the various partners, and the use of transgenic technology. Concretely, this means that the social negotiation process with all stakeholders involved not only focuses on the effectiveness of the host plant resistance, but also on the sustainability of the solution, the differentiation of socio-economic benefits that arise from the project, and the credibility of the safety testing and precautions that are undertaken. If these

[52] According to Indian legislation, approval for commercialization must be given for every transgenic event, rather than for a specific construct or transgene. This means that the required biosafety tests must be conducted on the final elite event material that will actually be commercialized on the Indian seed market.

aspects of the project are not addressed satisfactorily *as integral part of the project*, support for the technological development collapses, and the project would fail.

The technical design of a gene construct

As mentioned above, one of the key objectives of the project was the effective and sustainable resistance of the Brassica to diamondback moth infestation, both for large scale, as well as small scale producers. The proposed design with a dual Bt gene construct is not only expected to provide effective host plant resistance, but is also specifically designed to delay the build up of insect resistance against the Bt toxin. Current management of resistance build up in target insects comes down to the compulsory sowing of refuge areas with non-resistant crops. Next to the fact that this practice is difficult to control or enforce on a large scale, the requirement of sowing refugias is especially problematic for small scale farmers who need their entire holdings to raise some revenues from their farms. Current refugia standards require a minimum of 20% non-transgenic crop to be sown, and sacrificing such a significant portion of the already small holdings is economically unattractive and hence unpopular.[53] If the dual Bt gene construct is indeed providing a much more sustainable insect resistance, this means that the need for refugias evaporates, and that the technology loses its bias for large scale farmers, and becomes 'scale-neutral'.

Another remarkable aspect of the technical design relates to the precise design of the dual Bt gene construct. There are various ways of stacking multiple genes in transgenic crops, like iterative transformations, co-transformations with multiple gene constructs at the same time, or transformations of constructs in which multiple genes are linked (Halpin 2005). The CIMBAA consortium has chosen to work with a linked dual Bt gene construct which prevents future segregation of the two genes in later crossings with other wild material. Separating both Bt genes would seriously compromise the long term host plant resistance to the pest insect. While working with linked gene constructs generally provides a robust methodology for producing transgenics with stacked genes, it is technically more challenging than performing multiple transformations. Remarkable is that the best known other example of dual Bt resistance – Monsanto's Bollguard® II cotton – does not use the linked genes methodology, but is produced by successive transformations of two Bt genes.[54]

---

[53] It is sometimes claimed that resistance build up by the insect in an Indian situation with small holdings and a wide range of other wild, non-transgenic host plants for the insect, is not likely to occur very easily and that refugia are therefore in fact unnecessary (interview data). However, this point remains controversial, and the currently prevailing formal regulations that go along with conventional transgenic host-plant resistance are in fact strongly biased towards large scale farmers.

[54] According to information available on http://www.agbios.com/dbase.php?action=ShowProd&data=15985 (last accessed 17 September 2008).

Respondents from the CIMBAA project indicate that the dual Bt gene construct with linked genes is specifically designed for a context in which further breeding with the transgenic material by other breeders was expected, and where this should not cause any problems in terms of reduced host plant resistance (because of segregation of both Bt genes during crossing). While further breeding is also possible with non-linked dual Bt plants, the specific context and goals of the CIMBAA project did materialize in the choice for a dual Bt gene with two closely linked genes.

Hybrids or open-pollinated varieties; an example of contrasting requirements

Next to the specific gene construct used in transgenics, also the 'genetic background' in which this construct is being used is contested ground, and therefore a potential arena for redesign. Seeds can be produced as 'hybrids' or 'open-pollinated varieties' (OPVs), with different implications for farming practices. In Chapter 3 the development of maize hybrids had been introduced, as well as their implications for the use of farm saved seeds. Summarizing the argument once more: hybrids seeds are derived from a cross with two homozygous parental lines, and may benefit from heterosis, or hybrid vigour, which is claimed to significantly increase yields as compared to non-hybrid varieties. Essential is that the hybrids of specific parents (which is the generation that is marketed) are heterozygous. Although technically fertile, these hybrids create a strong segregation of agronomically valuable traits in a next generation, rendering their offspring much less attractive for commercial cultivation. In practice, this has meant that farmers are strongly encouraged to buy new seed every year, instead of reusing their farm saved seed for several years. Especially in cereals this has made an important difference, since the harvested produce has a dual character as both grain (to be sold or consumed) and seed (to be used for replanting next year).

In practice, the importance of farm saved seed is much less prominent in horticultural production (like cultivating cabbages), since the harvested produce is usually not simultaneously useable as seed. Moreover, Brassica seed production by farmers is generally not remunerative for the farmer, since it requires the crops to be on the land for the entire year, where the production for consumable vegetables allows two to three harvests per year.[55] Saving a part of the holding for seed production is then not very attractive. Moreover, most Brassica do not produce seed at all in warmer regions in lowland areas, because of the biological constitution of the plant, and its original adaptation to colder areas. For that reason, most Brassica seed production in India is carried out in the Northern highland areas near the Himalayas. This seed is then brought to southern regions for vegetable production near big cities like Delhi, with large market potential.

[55] Interview data.

Regardless of the generally perceived unattractiveness of farm saved seed in Brassica, the public private consortium has considered allowing the release of open pollinated material, which would potentially allow farmers to produce and reuse their own seeds. However, considering the transgenic nature of the crop, this would lead to concerns regarding crossing of the transgenic material with wild relatives. In practice, this has made the release of transgenic open pollinated varieties unlikely under the current biosafety frameworks. In fact, it remains open for the public partners in the consortium to decide how the sub-licensees will be allowed to release the transgenic material via their own varieties. However, the private seed company Nunhems has made explicit that it will commercialize the material as cytoplasmic male sterile (CMS) hybrids which produces infertile pollen, which in turn strongly inhibits the chance of any outcrossing with wild material. It seems likely that other companies will follow suit.

Two different concerns regarding the application of transgenic crops lead to conflicting recommendations here. Biosafety concerns limit the room for manoeuvre to allow farmers to save and replant their own transgenic seed. In the current understanding of potential risks that transgenic technology entails, it requires strict biological control, which cannot be guaranteed when using open-pollinated varieties. The project may not have been able to challenge the prevailing biosafety regulations and their implications, but what makes the case study interesting is that the technology developers did challenge the dominance of hybrids in vegetable production and attempted to create room for alternative production styles that might have explored the possibilities for farm-saved seed. The final decision to exclusively release hybrid seeds reflects the project's goal to present a transgenic crop that would cause the least controversy. Treating biosafety concerns as the first priority reflects that goal.

Tailoring the mode of commercialization

Interestingly, in this project, the redesign of the technology is not restricted to adapting the specific technological object (a transgenic crop) to the requirements of the project. Also the mode of commercialization is specifically adapted to the socio-political context in which the project is supposed to work.

One of the potential controversial aspects of transgenic seed is its relatively high cost, as compared to conventional seed. Usually, a private sector company receives a return on its investments in R&D by raising a technology fee. This fee may be calculated into the seed price, or may be paid as a royalty by other seed production companies that have sublicensed the technology from its developer. These companies in turn increase their seed price to recover the royalties they have to pay. The high investments associated with transgenics development and their deregulation (passing all the regulatory requirements for commercialization) usually imply a significant technology fee, and therefore a high seed price. This aspect of transgenic seeds has been criticized as being inappropriate for resource poor farmers, especially in combination with the hybrid character of the seed, requiring farmers to buy new seed every year. A significant increase in productivity, or savings on pesticide usage may be expected to easily compensate for a higher seed price. Nonetheless, the higher seed cost is generally perceived as an increased risk in cultivation, especially when the crop is grown in difficult conditions and harvests may fail.[56] In addition to the high seed price, the technology developer commonly gains a market monopoly on his patented technological product. This again is commonly criticized as a mechanism by which a company can exploit its position as patent owner, and push up the prices to maximize the benefits for the company, and to squeeze the benefits of other stakeholders in the production chain.

In order to make the technological project successful and to gain support from as many stakeholders as possible, the patent owner in the CIMBAA consortium (Bayer CropScience) has made the rather remarkable choice to make the Bt technology available, *without raising a technology fee*, thereby ensuring a low and affordable seed price.[57] Moreover, once the dual Bt gene construct has been developed and transformed into suitable germplasm, the ownership of the technology and biological material will be transferred from the company to a public partner in the consortium (most likely AVRDC). This institute will subsequently sublicense the technology back to Nunhems, and a range of other seed companies for the concrete breeding of the material into elite germplasm, for which every seed company may choose their own lines and niche markets. This means that not only the material and technology is released without a technology fee (that in the end farmers would have to pay), but also that the commercialization is not exclusively restricted to Nunhems Seeds, but instead open to any company or institute that might be interested. This implies a remarkable divergence from the ownership structure that was typical for conventional biotechnology development.

The relatively high price of transgenic seeds is usually legitimized with reference to the high level of investments associated with transgenics development, and the expected benefits for

[56] This has been one of the controversial issues in the commercialization of Bt cotton in India, the only transgenic crop currently released in the country. The high technology fee raised for the Bt technology, developed by the Monsanto-Mahyco Biotechnology (MMB) joint venture, led to a high seed price. Although the precise economic revenues for farmers are still an ongoing matter of debate, it is argued that the economics of the expensive Bt seeds do not work on marginal (rain-fed) land where the cotton yields remain low (Sahai 2007). Considering the uncertainty of good yields on marginal lands, high input costs constitute a serious danger for farmers that are not supported by insurance systems, but have to take loans against unfavourable conditions. See also Snapp *et al.* (2003) who argue that cash investments in production under fluctuating market prices may present a too high risk for many farmers, in spite of high yield responses.

[57] As for several other aspects of this project, it is at the time of publication uncertain how exactly this will be operationalized in practice. While clear intentions are being expressed that no technology fee should be raised, legislation on intellectual property, licensing and gifting may provide serious problems in making such intentions concrete. It will be interesting to follow what will remain of these intentions by the time a concrete product will be released on the market, but regardless of future outcomes, the sincere intention to commercialize Bt Brassica without raising a technology fee is real and worthy of investigation and explanation.

farmers in terms of increased productivity (offsetting the high seed price). The availability of public funds in the CIMBAA project already allows a significant reduction in any potential technology fee, since it reduces the level of investments that are required by the private partner. However, the company can also follow this strategy because it treats the innovation of transgenic seed in a different way than many biotechnology companies have conventionally done. In fact, the strategy that is being followed fits a company that is primarily a seed breeding company, rather than a biotechnology company.[58] Interviewed stakeholders from Nunhems Seeds claim that the genetic modification technology is merely instrumental in allowing for the continued cultivation of Brassica in India, which in turn allows the company to release new hybrids in the future. The added value for the company may primarily be in preserving a valuable market for new crop varieties/hybrids, rather than in licensing out new technologies. This implies that instead of gaining a position in the market based upon the ownership of a specific technology, what becomes more important is the value of an entire portfolio of traits, captured in the elite germplasm of the company. Moreover, Nunhems Seeds in India has advanced capacities in seed coating, providing the seed with fertilizers and fungicides. Currently, Nunhems Seeds already has a strong position in the sales of Brassica seeds in India. Rather than claiming a bigger portion of the market, the company has a strong interest in maintaining the market itself, or possibly enlarging it. Collaborating in a non-exclusive manner in a Bt Brassica project which has the potential to secure and enlarge the production of these cabbages in large areas of India is very interesting in that respect.[59] For the business model of the company, this means that instead of relying on one crucial trait, and the protected ownership of that trait, the company relies on a whole package of superior germplasm, seed coating technologies, and a good network of distributors and buyers to secure its position in the Indian seed market.

An additional important consideration that has played a role in following this strategy is the large size of the Indian seed market, which allows the company to invest in this technology development project, and to gain return on investment through prolonged and large scale seed sales in the country. Moreover, considering that the country is too big and diverse for one company to serve all Indian vegetable farmers, there is little harm in allowing other companies to serve specific niches within the Indian seed market. In fact, since one seed company would have great difficulties in catering for the diverse needs of the entire country. Keeping the entire market closed would probably lead to the spread of seeds fit for one agro-climatic region to another, where they would perform poorly. This would in turn damage the reputation of the transgenic seeds, as well as the reputation of their producer. Finally, keeping

[58] This presumes that the initiative of this project has been with Nunhems, rather than with the actual patent owner Bayer CropScience, which is indeed supported by interview data.

[59] Note that cabbage cultivation in higher and colder regions of India is no problem, since insect attacks can relatively effectively be managed there with conventional IPM approaches. However, the very lucrative markets around big cities like Delhi and further down south require technological intervention to remain economically feasible in the long run.

the price of official and certified seed low may be the strongest weapon in preventing the large scale spread of uncertified seed, which is currently a significant problem for Bt cotton in India (Jayaraman 2001, 2004).

The significance of this aspect of the CIMBAA project is not primarily in the low price of the transgenic seed, which makes the product appropriate for resource poor Indian farmers, in practical terms. More interestingly is that while the introduction of transgenic host-plant resistance has conventionally also implied the need to get a return on high investments in the technology development, by establishing a market monopoly by the patent owner, this consortium is adopting a model of commercialization that changes that social organisation. This can to some extent be explained in terms of the aims and interests of the projects, building credibility and legitimacy, and with reference to a number of specific enabling conditions. Nonetheless, the important conclusion must be that redesign of the technology is not only possible in instrumental terms, but also in terms of social relations that are implicitly established in the production system. As will be discussed by the end of this chapter, this has some implications for the way in which the technology developer positions him/herself in the production system, and hence deals with the mechanism of appropriationism.

Representation of farmers, rather than participation

The analysis of the CIMBAA case study so far has demonstrated that the project has responded to publicly expressed concerns over transgenic technologies in a number of ways. This both influenced the material design of the transgenic crop, as well as the strategy for commercialization. However, one question that has not been answered is how the project leaders knew or assessed what concerns to respond to. Did they find a way of involving various stakeholders or end-users in the process? And has this led to a careful attuning of new technological solutions with existing practices and expertise?

The CIMBAA project is very conscious of the controversies around the development of GM crops, and is clearly aiming to come up with a technical solution for the pest problem of Indian farmers, that will deal with the criticisms levelled at transgenic technology. With that in mind, it is remarkable how limited the role is that farmers and civil society organisations have played in the CIMBAA project, especially in its early phases. Stakeholder involvement has frequently been mentioned as a way of improving research relevance, and facilitating a wider adoption of new technologies. This also emphatically applies to relatively upstream participatory methodologies, that rigorously document farmer preferences and involve them in decision making (Snapp *et al.* 2003; Snapp and Heong 2003). Sperling *et al.* (2001) mention examples of successful participatory approaches like surveys and focus groups to document farmer knowledge, use of expert farmer panels, and incorporation of farmer views into criteria used to rate performance. Such methodologies for the involvement of farmers and other stakeholders would appear attractive for the CIMBAA project to make sure that all the efforts to produce something useful and attractive will indeed result in a transgenic crop that will be widely accepted.

Moreover, it is important to stress that even at the initial stages of technical design of a potential solution to the diamondback moth infestation, a number of things are already at stake that reach beyond merely technical issues. In the early phases of the project, very specific choices have been made about what to address in the technical design of the transgenic crop, and what not. While effective insect resistance and delay of resistance build up by the insect were chosen as major scientific challenges, to be solved by a technical design (hence the dual Bt construct), a number of other known issues were treated as 'management issues'. These included the extent of outcrossing to wild relatives, and issues of co-existence of transgenic and non-transgenic crops, which were labelled as issues to be taken up in further research (to what extent does outcrossing take place, and is that a problem?), or as management issues to be decided upon at a political level (is co-existence and labelling of transgenic and non-transgenic crops feasible in India, and if not, is that a problem?).

The prioritization of what counts as essential objectives to be solved by a technical solution is not self-evident, but depends on the specific understanding of the problem at hand. Other stakeholders, like for example organic farmers, may have stressed to address the issue of co-existence of GM and organic farming methods, or the tracing and labelling of the GM products *as integral part of the proposed technical solution.* By treating the design of the transgenic crop as a technical, non-political issue, a specific but arbitrary perspective on the problem is privileged over alternative views. This understanding of the relation between technical design and socio-political relevance of the end product argues for an involvement of a wider range of stakeholders in early phases of discussing the problem and potential solutions, in order to avoid an externally imposed framework of how problems with pest insects in vegetable production in India should be solved.

The CIMBAA project has undertaken a number of initiatives to get in touch with the local perspective on agricultural development, and the potential role that a transgenic pest-resistant cabbage could play. For example, a socio-economic study has investigated the pesticide usage and the potential interest in a transgenic solution among farmers (Sandur 2004). Also, one farmer representative has been present at a launching workshop in February 2005 to express his support for the project (Srinivasan *et al.* 2005). In addition, the intention is to work with farmers in developing IPM practices once the transgenic material has been developed and approved for field trials. Finally, a number of NGOs had in fact been invited to the launching workshop in 2005, but (according to respondents from CIMBAA) turned down the invitation. However, while interaction with farmers is likely to increase in the later phases of the project, all in all these initiatives have been rather modest in involving both farmers and critical Indian NGOs in early stages of technology development.

In spite of the importance of stakeholder involvement in early stages of the project, a number of pragmatic problems generally arise in setting up stakeholder involvement in technology design. While priorities can obviously be set in a participatory way, the lack of specialized technical expertise disqualifies outsiders from playing an active role in designing e.g. gene constructs. Moreover, difficult decisions need to be made regarding who to involve, and where to draw the boundaries of involving 'outsiders'. But more importantly, CIMBAA project leaders stress the vulnerability they perceive with respect to criticism from outsiders on the project and the proposed solution. Avoiding a too dominant position in the final market commercialization is one of the concrete attempts in order to prevent too much controversy being attracted to the private company involved. Moreover, communication about the project and potential challenges ahead is carried out with great care. It is reported that the project does not want to get out of the way of discussions with critical NGOs, but these discussions should be taking place at a moment that concrete testing data about the actual host plant resistance, yield, and biosafety are available. However, this is obviously in contrast with a very open and deliberative process of technological innovation. In other words, a paradox emerges when 'society' calls for an open and transparent innovation process in which all relevant stakeholders can have a say, but at the same time a context is created in which it is very difficult for such technological projects to actually adopt such an approach in early phases of the project.

While the CIMBAA project has clear aims in both addressing the needs of specific farmer groups in India, as well as in building legitimacy for their technological product, the context here described leads to a 'representation' of farmers and their needs, instead of their direct 'participation' in the project. The project has a strong stake in being demand driven, and as such the project is very consciously aware of its final end-users and other impacted groups. Moreover, the public nature of the CIMBAA project demands that the project not only demonstrates a good final end-product, but also shows an open and responsive attitude during the innovation process itself. However, there are strong disincentives for the public private consortium to adopt such an approach. In practice this means that the technology is tailored to an 'image' of the Indian cabbage farmer which is produced within the consortium itself, rather than in direct dialogue with farmers or other stakeholders themselves.

This specific observation regarding the degree and mode of stakeholder involvement in the CIMBAA case raises the question to what extent this experience is representative such technology development projects in general. The case study here presented did not find any objections of principle that would withhold stakeholders from playing a bigger role in technology development projects. What the case study did observe was a set of specific disincentives for the project to actively involve farmers in the early stages of technology development. These disincentives were related to the importance of biosafety regulation in this project and the vulnerability in terms of public opinion, which were in turn a direct consequence of the project's work on transgenic crops. For this reason, it is to be expected that in other projects working on transgenic crops, similar disincentives may occur.[60] This leads to the conclusion that although stakeholder involvement may in principle play an important and useful role in such technology development projects, the specific choice to use transgenic technology makes direct democratic control by impacted stakeholders at least very difficult, in spite of the explicit aim of the project to be demand driven and responsive to farmers' needs.

The role of intellectual property and liability

The discussion in the previous section already illustrated that although the consortium makes various attempts to make transgenic technology appropriate for small scale vegetable farmers, this does not lead to a very open and interactive innovation process. This section will focus on the importance of intellectual property and biosafety regulations on the setup of the consortium and on the central role of the seed company as main source of patented technology.

Intellectual property and freedom to operate

The role of the private sector company and its ownership of (or unrestricted access to) the relevant intellectual property has been an essential condition in setting up the consortium as it is and in realising its acclaimed intentions. One of the major challenges in developing and commercializing transgenics in collaboration with other institutes is making arrangements for the sharing of intellectual property. Bt technology for insect resistance is one of the few traits (next to herbicide tolerance) that has been widely applied and commercialized in GM crops worldwide (James 2006). As a result, the technology is subjected to a great number of patent claims. Although it is difficult to get a precise overview over the reach of a patent portfolio, the private sector company trusts that it will have access to all relevant intellectual property (IP) through the patent portfolio of Bayer.[61] This should allow the project to develop and commercialize the Bt Brassica, without being dependent on negotiations with other companies over licensing of their patents. Importantly, CIMBAA project leaders consider this guaranteed freedom to operate to be more than just convenient, but as practically essential for the project to succeed. Their concern does not primarily relate to the development phase,

[60] In fact, research by Glover (2007) confirms that the observed mechanism of 'representation' rather than 'involvement' is not unique for this project. He describes how the Monsanto Smallholder Programme also fails to substantially involve the stakeholders it claims to focus at, albeit for a different set of reasons.

[61] Interviewed stakeholders mention a number of around 20 patents that apply for the development of Bt Brassica by CIMBAA. However, the precise number or their accessibility is hard to determine, for example because some patent claims are currently being contested. See also Graff *et al.* (2004) who explain that even for patented technologies it is often not entirely clear who owns what. This uncertainty can be cleared up in courts through patent interference cases, but these cases can drag on for years. Moreover, for most registered patents, there is no such scrutiny. They conclude that *"as a result, the boundaries for a considerable expanse of technological territory are not clearly demarcated, creating considerable uncertainty as to when a new application could be considered to be infringing or trespassing"* (Graff *et al.* 2004, p. 124).

but the phases of deregulation and commercialization for which it is much more difficult to get access to patented technologies or traits. Note that this is in general a problem with the development of transgenic crops, but that the transaction costs in negotiating freedom to operate are generally relatively higher for horticultural crops because of their relatively small market (as compared to major crops such as maize and soybean) in which any transaction costs have to be recovered (Graff *et al.* 2004).

As has been discussed in Chapter 3, the issue of limited access to protected intellectual property is potentially very problematic. Any program developing biotechnologies has to negotiate its access to potentially protected technologies and traits, whether it is about private sector product development, public sector development project, or a combination of both. In this playing field, companies with extensive patent portfolios can use their patents as bargaining chips in negotiating access to the intellectual property (IP) of another company. Moreover, they often have the financial means to buy their way to patents they need (either by paying royalties, or by simply acquiring another company with valuable and interesting patents). For public sector programmes, getting a 'research exemption' for performing research on patented technology or traits may not be a problem. However, commercialization of the developed material is often not allowed, without the payment of additional royalties to the original patent owner. Having one private partner with access to all relevant IP is a great advantage because it saves a lot of time and effort in negotiating transfer agreements and potential royalties. In fact, while the option of inviting more private partners to the consortium has been discussed, the decision has been made to stick to one private partner in order to keep things simple and workable.

Biosafety and liability

Next to these complications raised by intellectual property rights on key technologies, biosafety regulations constitute an important background for projects working on transgenic technology. This complexity is further exacerbated because of a direct relation between the intellectual property of specific technologies or traits, and the liability in case anything goes wrong with a product in which these technologies or traits are being used.

The 1992 Convention on Biodiversity (CBD) was accompanied by the Cartagena Protocol on biosafety, which came into force in 2003, and which provided an international framework for the handling of genetically modified organisms. However, the Cartagena Protocol and the CBD left an important point for debate open to resolve among the member states of the CBD, which was to draw up a framework for liability and redress regarding the use and spread of

Living Modified Organisms (LMOs).[62] Meanwhile, some national governments have adopted national liability regimes, which hold a broad view over who may be held liable. For example, the Nigeria Biosafety Guidelines state that *"liability shall attach to the applicant, the person responsible for the activity, which results in the damage, injury or loss, as to the provider, supplier or developer of the LMOs/GMO(s) of products thereof."*[63] India also has a number of biosafety regulations in force, but is currently still in the process of developing a national liability and redress framework, in line with the negotiations over international regulations.[64] With respect to the adoption of a liability and redress regime at the international level, Sullivan notes a trend in arguing that liability for GMO-related harm should be imposed directly on the holders of patents that cover a GMO (Sullivan 2005).

This dynamic regulatory context has a very direct influence on how the CIMBAA consortium deals with transfer of technology and transfer of liability, as well as more indirect influences on the type of collaborative organisations that are preferred. Stakeholders from both public and private partners in the consortium report their high confidence in the safety of Bt technology and the reliability of the safety testing that will be done prior to commercialization in India. Nonetheless, even in this rather orderly and relatively simple situation with one major patent holder, liability is an issue, considering that the transgenic material will be licensed out to potentially many other public and private sector institutes that will commercialize transgenic Bt Brassica seeds. If the commercial release of the crop in India is allowed under certain conditions regarding spacing to non-GM crops, labelling, or combinations with IPM strategies, the failure of other parties to comply with these measures may lead to future litigation. In order to prevent such situations to result in liability claims against Nunhems/Bayer as original patent holders, efforts are being made to investigate to what extent the transfer of ownership of the technology to AVRDC (or another public partner in the consortium), can also imply a transfer of liability. While this may not relieve the private company fully of the risk of future liability claims, it should significantly reduce the chance.

These ongoing debates and negotiations about liability and redress regimes at both the international and national level have their implications for the setup of public private collaborations in developing biotechnologies for developing countries. Sullivan notes that *"one can imagine disincentives for private enterprises to donate GM technology for humanitarian uses that would arise from a rule imposing liability on patent holders and developers of GMO in*

[62] Article 27 of the Cartagena protocol. The conference of the parties agreed to complete the process of negotiating a framework of international rules and procedures related to liability and redress within 4 years from the protocol coming into force on September 11th, 2003. However, at the time of writing a final text for this framework had not yet been agreed upon by the 'open-ended ad hoc working group'. See http://www.cbd.int/biosafety/issues/liability.shtml (last accessed 17 September 2008).

[63] http://bch.biodiv.org/doc/leg/nigeria_biosafety_guidelines_2001.pdf (last accessed 17 September 2008). Quoted in Sullivan (2005).

[64] See Chaturvedi and Chawii (2005) for an overview of recent Indian Biosafety regulations.

all cases." (Sullivan 2005).[65] This may obviously complicate matters for public private consortia that are developing GMOs for humanitarian purposes (even when commercial incentives are at stake as well). Licensing essential technologies from multiple patent holders, potentially under humanitarian use licences[66] would require all these patent holders to trust the safety of the project, and to be willing to take the risk of future liability issues which may in fact come their way, as they are the original developers of the technologies.

Implications for the structure of public private consortia

The combination of the proprietary nature of technologies and traits, and the potential liability of the owner in case of breaches of biosafety, can make access to technologies and traits very difficult indeed. The direct implications for innovative biotechnology development with a public or humanitarian function are that it may be very difficult to materialize creative ideas to develop pro-poor biotechnologies that serve a public function. The CIMBAA project overcomes these limitations, by partnering with a big life science corporation with an extensive patent portfolio. In fact, in the CIMBAA project, the match between what was indicated as the 'most ideal technical solution' for the insect problem and the intellectual property that was available to the private partner in the consortium, turned out to be very good. While it may not be impossible to partner with multiple companies with smaller patent portfolios, the current regulatory context in terms of intellectual property rights and biosafety frameworks introduces a strong preference for collaboration with only one private partner, and one that has access to an extensive patent portfolio.

This means that while a non-exclusive model for commercialization of the Bt Brassica will be adopted in the CIMBAA project, the case study simultaneously illustrates how the position of biotechnology companies and seed breeding companies in fact remains very central. In this specific project, the ownership of key patents may no longer be used to enforce a market monopoly on the level of seed sales, the exclusive ownership over the Bt patents does structure the setup of the public private partnership in a way that very much legitimates the central position of the owner of the relevant intellectual property. The fact that the Bt technology is made available under a humanitarian use licence for this project does not change that observation.

[65] Louwaars (2005) has voiced a similar concern.

[66] See Chapter 3, the section on *'Challenging the trend towards limited access to technologies and new varieties'*, Brewster (2005), or Louwaars (2007, p. 129) for an elaboration of the concept of a humanitarian use license.

Discussion – Farmers as recipients of technology

Evaluating a project like CIMBAA is difficult and tricky, because it strongly depends on the perspective that is taken and the questions that are asked. What is a public private consortium like this supposed to respond to? What is it supposed to deliver? Like stated in the introduction to this chapter, this case study is not evaluating the project in terms of successfulness of the final technological solution, nor making an argument pro or against the use of GM crops. Instead, the case study has been motivated by an interest in how a project like CIMBAA interprets and operationalizes what it means to deliver an 'appropriate' technology for resource poor farmers in India, and to what kind of adaptations and reconstructions that may lead. This has led to a primarily descriptive and reflexive presentation and discussion of the project, rather than to a discussion of concrete ways to improve the project in a practical sense. Nonetheless, the discussion does indicate in what ways the project can be very flexible and oriented at the problems of resource poor Indian vegetable farmers, and in what ways the project is caught up in structural trends that are much harder to challenge or change.

Appropriateness in the CIMBAA consortium

For CIMBAA, appropriateness is defined in different ways, some more explicit than others. In terms of the technical functioning of the technology, it is crucial that the transgenic cabbage has to exhibit effective insect resistance in a way that does not lead to a quick resistance of the Diamondback Moth to the Bt endotoxin. In other words, whatever technology is developed, it should be able to deal with the problem of insect infestation in cabbage production. Moreover, the final product has to be available for, and work economically for resource poor Indian vegetable farmers. This puts limits on the pricing of the seed of the new cabbage variety, and on the production with or without non-transgenic refugia. These criteria are translated both into the design of the dual Bt gene construct, as well as in the non-exclusive strategy for commercialization of the cabbage seed. While the dual Bt gene construct – at least theoretically – eliminates the need for refugia, the non-exclusive commercialization strategy allows for a differentiation in the market in which Nunhems can choose to develop high quality hybrids for farmers who can afford them, while other companies or public sector institutes can develop cheaper varieties with the same insect resistance, but not using Nunhems elite germplasm as 'genetic background', nor the advanced seed production and coating facilities. This market differentiation is expected to make the technology available for small scale and resource poor farmers as well.

These criteria for appropriateness for CIMBAA mainly focus on the *product* that is developed. The transgenic cabbage should be effectively insect resistant, cheap, safe, and commercially interesting for resource poor, smallholders. In addition, some attention goes out to the appropriateness of the *process* of innovation. For example, in order to increase the credibility of the safety testing, this part of the project is emphatically outsourced to the public sector institutes in the consortium. This way, the consortium aims to avoid any accusations that safety testing by a private company is biased because of the commercial interests at stake. Secondly, the release and commercialization of the transgenic cabbages is planned to be accompanied by a 'stewardship' programme which is supposed to monitor resistance build up by the insect, and continue working on new generations of the insect resistant crop. This means that the innovation does not end with a product entering the market, but is conceptualized as an ongoing process of monitoring and renewal.

But especially in terms of the innovation as 'process', some questions emerge. First of all, in order to increase the enthusiasm and acceptance of the technology by farmers, the project is argued to be demand driven, i.e. responding to a clearly perceived need for insect resistant cabbages by Indian farmers. This may be true, since it does take a widely perceived problem in cabbage cultivation as starting point, but then it externalizes the process of defining what exactly is the problem, and what trajectories can be followed to address it. In fact, while the project undeniably contributes a potential solution to the DBM problem in India, the history of its creation suggests that it is much more inspired by the perceived potential of Bt technology – and hence technology-driven – than informed by the perceptions of farmers, consumers and civil society organisations in India. In other words, the project may be demand driven, but it still is rather top-down in its approach. This perception of the consortium is further confirmed by the limited scope of stakeholder involvement in the early stages of the project.

The consequence is that the transgenic cabbages developed by CIMBAA may respond to the insect problem in a technically effective way, but one that fits the perspective of an external seed supplier like Nunhems. That is not to deny that a win-win situation can exist where farmers are helped in their cultivation, and Nunhems secures an interesting Indian seed market. However, this approach to finding a solution for insect infestation does potentially exclude a wider range of approaches that might have been possible. Whether alternative approaches to solving the DBM problem are possible is in fact a point of debate. CIMBAA project leaders legitimate the project by arguing that integrated pest management approaches have been tried, and have failed to control the DBM infestation in cabbages. This would mean that apart from stopping cabbage cultivation outside of the highland areas of India, the only solutions are the extensive spraying of pesticides, or the development of host plant resistance. Indian NGOs working in agriculture and developing non-pesticidal cultivation methods challenge that representation of the problem, and in interviews have argued that a diversification in production and the use of organic pest control methods cán in fact provide successful ways of managing insect infestation.

The question whether alternative solutions to transgenic host-plant resistance would have been possible and feasible goes beyond the scope of this chapter and this research. Moreover, even if a consensus would be reached that the transgenic cabbages are technically the most ideal solution, care should be taken not to let the outcome legitimize the process. The main point of interest for this thesis is the way in which the project has operationalized the notion of 'appropriate development', and what this means in terms of restructuring social relations of production and innovation. As demonstrated, the approach adopted by CIMBAA strongly focuses on providing an excellent technical solution, but does so by taking the technology development in house and leaving local stakeholders with a very limited role in the process. This at least applies to the development of the transgenic crop. In later phases of the project, the innovation process is likely to become more interactive, when the crop is evaluated in an integrated pest management setting. This is not only expected to make any type of host plant resistance less vulnerable to insect resistance build up, it also implicitly acknowledges that there are multiple points of innovation in process of solving the insect infestation, which require the innovative capacity of farmers and agronomists, next to the products of a seed breeding company.

Supporting the prominent position of the technology developer

The case study has demonstrated that the project consolidates rather than challenges the central position of the seed company in the future production of vegetables in India. The release of open-pollinated varieties instead of hybrid seeds has been discussed, but dismissed. It should be noted that the focus on a vegetable crop, rather than e.g. cereals is of relevance here. Like aforementioned, it is highly uncommon for Indian cabbage farmers to produce their own seed, meaning that the project does not lead to a further externalization of the seed supply, but merely consolidates a situation that was already there. However, while the external provision of vegetable seed is not new, the introduction of transgenics in this vegetable production system, with the accompanying biosafety regulations and technical control, does have an interesting repercussion for the strategic position of Nunhems. For example, the requirement to only release cytoplasmatic male sterile hybrid seed is relatively advantageous for seed companies – like Nunhems – with strong technical capacities, as compared to their competitors without such facilities. This means that the project does not shift seed production from farmers to a seed company (which has already happened in the past with the commercialization of vegetable production), but it does introduce a bias for the production of seed by companies with rather advanced technical facilities. In terms of a mechanism of 'appropriationism' – as discussed in Chapter 3 – these observations imply that no further commodification of a previously public good takes place, but that the material design of the transgenic and CMS hybrid seed does allow the seed developer to claim a prominent position in the future vegetable production system. In other words, it highlights the social meaning of the technological design in terms of the relative importance of the seed developer in the vegetable production system.

Finally, the CIMBAA case study has demonstrated the crucial role of intellectual property arrangements, although the precise evaluation of this case study with respect to a trend of increasingly strict intellectual property protection depends strongly on the questions asked. On the one hand, the public private partnership between Nunhems/Bayer and the public research institutions has meant that proprietary technology has become available for this project, and hence for the development of plant biotechnology for resource poor farmers. This may be taken as an example of how public private partnerships can break open the sometimes clogged patent landscape, and facilitate the transfer of technology from multinational biotechnology companies, to resource poor farmers. At the same time, the ownership structure itself has not been challenged by this project, as proprietary technology has become available for the CIMBAA project, but has not been donated to the public domain. This means that the commercialization of transgenic cabbages outside of India, or the use of the same technology in other crops, can still be restricted by the patent holder. Moreover, the importance of the patent holder has been confirmed with respect to the institutional setup of the CIMBAA consortium. The observation that negotiating licences and royalties with more than one patent owner would greatly complicate matters for the consortium has been a crucial reason to involve Nunhems seeds as the only private partner in the consortium. This reconfirms the central and powerful position of biotechnology companies with a large patent portfolio.

In conclusion

The chapter discussed how adaptation of transgenic technology has taken place to make the technology appropriate for the specific objectives of the CIMBAA consortium. This has led both to rather straightforward instrumental adaptations, as well as to a more profound reconstruction of the social meaning of transgenic crops by making it more effectively available for small-scale Indian vegetable farmers. However, at the same time, the agricultural development spurred by the CIMBAA project remains a largely externally designed affair which is determined by an indirect representation of Indian vegetable farmers and critical NGOs, rather than by their direct involvement. This demonstrates how the project is to a large extent determined by the desire to embed the vegetable seed company in the production system, although in a new and innovative way. The upcoming two chapters will discuss case studies with entirely different dynamics and rationales, in which the projects do move beyond an indirect representation of farmers and other local stakeholders, and more extensively aim for their direct involvement in agro-technology development.

Chapter 5

Reconsidering the role of potato farmers in breeding and multiplication – Experiences of the International Potato Centre

> *"Our business is to produce knowledge and new technologies. Probably the technology that will contribute most to the Millennium Development Goals are improved varieties. Given the size of our budget, and our recognition of the potential for impact, that would be it."*
>
> (anonymous CIP research manager, June 2007)

Introduction [67]

One of the points clearly illustrated by the previous chapter, discussing the work of the CIMBAA consortium, was that the objective to develop technology for resource poor smallholders may have repercussions for technical design, and conversely that technical design may reflect the priorities and strategies of a specific project. This chapter discusses a different case study in which plant breeding is also used and adapted to serve farming systems of resource poor farmers, but in which the dynamics of adaptation are quite different. Perhaps most crucially, the case study in this chapter demonstrates that making technological development appropriate does not necessarily require a profound technical redesign, but may strongly depend on the 'sociotechnical ensemble' of new crop varieties, breeding approaches, marketing strategies, and other ways in which a technology is embedded in a specific project for agricultural development.

Concretely, this chapter will focus on the work of the International Potato Centre (Centro Internacional de la Papa, CIP) in Peru, and it is attempts to develop improved potato varieties for potato farmers, both in the Peruvian Andes as well as in other parts of the world. In contrast to the CIMBAA case study presented in the previous chapter, the work of CIP is entirely

[67] This chapter is largely based upon material previously presented at Tailoring Biotechnologies conference: "Redesigning Agro-Biotechnologies for Development?", Kyoto, 3-5 November 2007. The accompanying paper is published as Vroom, W. (2008). 'International agricultural development as contested ground: Three levels of resistance and reconstruction.' In: Ruivenkamp, Hisano and Jongerden (eds.) *Reconstructing Biotechnologies*, Wageningen Academic Publishers, Wageningen.

publicly funded. This means that private sector interests or intellectual property issues do not play a significant role in this case study of the breeding work of CIP. Still, the institute does encounter a lot of difficulties that require it to make choices in its work, reconsider its role in the innovation system, and to set priorities.

One of the defining aspects of the work of CIP is that it is captured in a field of tension between breeding and releasing improved potato varieties, and a concern over the loss of traditional potato varieties that constitute an extremely rich source of potato genetic diversity. To put it bluntly: the more successful the adoption of newly released varieties, the quicker and completer the loss of these traditional varieties is expected to be. Interestingly, the loss of agricultural biodiversity is perceived as a problem from different perspectives, both from the perspective of an industrialisation of agriculture, as well as from the perspective of a non-industrializing agriculture. However, for the first, traditional varieties are primarily a source of genetic diversity for future breeding, while their cultivation in itself represents little added value. Therefore, the conservation of traditional varieties is crucial but can be done through collection and storage in seed banks, or through the ongoing cultivation of traditional varieties on small plots of land. However, from the perspective of a non-industrializing agriculture, crop genetic diversity can be a crucial characteristic of the farming system itself, and in addition represents an important culinary and cultural resource for indigenous communities. Therefore, rather than settling for a limited conservation of biodiversity, the question is to what extent it is possible to challenge the bias in breeding or agricultural development towards a homogenization of the farming system and the varieties that are being cultivated.

In summary, there may be a widely shared interest in maintaining or conserving agricultural biodiversity, but there are important differences in the reasons underlying this common goal, which lead to different approaches for the conservation of crop genetic diversity. This raises the question what kind of conservation strategies are being followed by CIP, and why. Concretely, the question arises whether and how the work of CIP reflects a specific approach to agricultural development, treating biodiversity as a commodity, or as living concept underpinning peasant based production systems.

In addition, the case study addresses the replacement of informal, farmers' seed systems by more formal and commercialized potato seed systems. Potato is a vegetatively propagated crop, meaning that farmers can use part of their potato harvest as 'seed potatoes' for the next season. This would make farmers and farmer seed systems potentially self-sufficient, if it were not for the build up of viruses in potatoes over the generations, and the subsequent 'degeneration of potatoes'. Dealing with virus infestation presents farmers and CIP with the dilemma whether to outsource the production of virus-free seed potatoes, or to look for ways to empower farmers in their own virus-free production of seed potatoes.

Finally, the chapter will briefly discuss how the reception and use of genetic technologies is not only determined by their technical function or material design, but also by the discourse that surrounds them. By discussing the controversy in the development of transgenic potatoes, and the representation of molecular fingerprint data in a remarkable way, the last section of the chapter demonstrates how CIP deals with the cultural connotations of new technologies, and in some cases is able to influence those connotations in a creative and innovative way.

In conclusion, the case study presented in this chapter does not focus on a single project, but on a set of initiatives developed by the institute that engage with the problems of potato production in different ways. Together, they provide a picture of a research institute that is struggling with the tensions in its work, and is looking for innovative and creative ways to combine the potential of modern plant breeding, with the richness of traditional potato varieties and seed systems. This requires adaptation and creativity in a variety of ways, but does not always require entirely different technologies. Instead, as will be discussed, essentially similar technologies may represent entirely different approaches to agricultural development, depending on the project and discourse in which they are embedded.

The International Potato Centre – Producing global public goods for potato farmers

The International Potato Centre (CIP) is one of the international public sector research centres of the Consultative Group on International Agricultural Research (CGIAR) that works on a specific set of mandate crops (potato, sweet potato and a collection of Andean roots and tubers) and attempts to develop 'global public goods' primarily through releasing improved breeding material. The notion of 'global public goods' is central to the work of all CGIAR institutes and implies first of all that the research outputs remain in the public domain, and are therefore freely accessible, and secondly that its research is in principle relevant for a wide range of contexts of application, rather than focusing on a very specific localized problem. The location of the centre in Peru –the centre of potato domestication and home to some 4,000 potato varieties (Spooner *et al.* 2005) – is by no means coincidental. However, the primary target areas and people for CIP are not necessarily in Peru or even South America. Potato and sweet potato are cultivated in large parts of Africa and Southeast Asia too, where the potential impact of improved varieties may be much bigger than in the highly diversified Andean potato systems.[68] Nonetheless, CIP greatly benefits from the natural potato biodiversity of the Andean region as a resource for their breeding work, and devotes significant attention to potato production in the Andean region in South America. The institute is involved in a number of programs to alleviate poverty and hunger among potato producers in the High

[68] See CIP's World Potato Atlas with maps of global production and average yields at http://research.cip.cgiar.org/confluence/display/wpa/Home (last accessed 17 September 2008).

Andes, as well as in projects aimed at *in situ* conservation of potato genetic diversity (e.g. CIP and CIDA 2006).[69]

Like other CGIAR breeding institutes, CIP has a specific position in the 'agricultural innovation system', which is relatively upstream but still clearly committed to doing 'research for development'. That means that the institute does not invest heavily in e.g. large scale genome sequencing programmes which are considered to be too far away from application in developing world potato farming. Instead, CIP is involved in genetic research to specific traits, like disease resistance, and is using that knowledge for pre-breeding programmes. The institute aims to develop biological material ('germplasm') that contains interesting traits for potato farmers. On a national level, CIP works together with national agricultural research institutes, like INIA (Instituto Nacional de Investigación Agraria; National Institute for Agricultural Research) in Peru. While CIP develops biological material with new and potentially interesting characteristics, it is these national institutes that have breeding programmes for specific varietal development, tailored to the conditions and preferences in the respective countries. While CIP provides new parental material for such downstream breeding programmes, they cross the new (resistance) traits into locally adapted germplasm.

For the development of potato germplasm with new resistances or other interesting agronomic characteristics, CIP relies on a range of traditional breeding techniques, as well as more advanced marker assisted selection methodologies. Research to the genetic mechanisms behind resistance traits are an important aspect of the work at CIP, for example involving the use of DNA array technology. Moreover, the institute has been experimenting with transgenic potatoes, and has especially been considering to introduce Bt insect resistance (Ghislain *et al.* 2003; Buijs *et al.* 2005). Concerns over outcrossing with wild relatives and negative public perception have made the institute decide not to release transgenic material in Peru, the centre of domestication for potatoes (CIP 2007). However, transgenic technology may still play an important role in future attempts to reduce insect infestation in potato and sweet potato production in Africa and Asia.

For CIP, the use of genetic technologies in breeding – whether transgenics or marker assisted selection methods – is an integral part of the wider breeding efforts of the institute. Molecular markers and other genetic technologies greatly facilitate the understanding of genetic mechanisms, and the selection for specific traits. Still, traditional breeding remains essential to select for useful parental material to release to varietal breeding programmes. The distinction between traditional breeding and molecular breeding therefore more or less dissolves in the

[69] *In situ* is Latin for 'in its place'. In the context of the conservation of genetic diversity, *in situ* conservation means that traditional varieties are conserved by their cultivation in farmers' fields or in designated parks, instead of collected and kept in seed banks (which would be away from the original context, and therefore *ex situ*).

work of CIP, and in its approaches. More relevant than the technological basis of new traits is the choice of traits that CIP breeding programmes work on, the precise understanding of resistance mechanisms in order to achieve sustainable host plant resistance against viruses, and the way in which new improved varieties will fit into current farmer production systems.

Because of the importance of matching new varieties with the preferences and production systems of farmers, the institute has a tradition of, and some renewed interest for involving various stakeholders, and notably farmers, in its breeding work (Thiele 2000; Thiele *et al.* 2001a,b). Moreover, it is investing in more reflective, sociological research into agricultural innovation systems, and agricultural knowledge and information systems (AKIS) (Ortiz 2006).

The challenge of developing appropriate new potato varieties

The question of what constitutes appropriate biotechnology in the work of CIP, is complex considering the wide diversity of projects that CIP is involved in, and the global perspective of the institute. CIP works for potato farmers in completely different regions in the world, who farm under different agro-climatic circumstances, with different production systems, and encounter different problems. The uptake of new varieties has traditionally been most successful by farmers with a strong market-orientation, who cultivate potatoes in relatively large scale and homogeneous plots, and are able to invest in some fertilizer and pesticides to protect and increase their production. This is the kind of production system in which the external supply of high-quality improved potato seed is remunerative and an approach of agricultural industrialisation generally is economically attractive for farmers. However, this is only a small subset of all potato farmers, especially in the Andean region of South America (Thiele 1999).

Peru is the centre of domestication of potatoes, and for that reason harbours an immense genetic diversity in potato landraces. The country has a long history in potato farming, which is argued to go back to 5,800 BC (Pickersgill and Heiser 1978; Weatherford 1988). The harsh conditions in the high Andes, in terms of weather conditions as well as the sometimes virtually vertical pastures, have traditionally led to a potato production system that is more than anything geared towards resilience and adaptation to variation (Halloy *et al.* 2005). In these traditional production systems, a high degree of agricultural biodiversity allows for adaptation of agricultural cultivation to the sometimes sharply contrasted micro-climates, and makes sure that at least a significant portion of the harvest survives climatic peculiarities of every year. Jan Douwe van der Ploeg described how this system of indigenous potato cultivation is like a 'craft', in which generations of experimenting farmers have been able to match the varying climatic conditions and soils with the variety in potato landraces, leading to a system that is exceptional in terms of productivity and sustainability, given the environmental conditions (Van der Ploeg 1993).

In his book 'The Botany of Desire', Michael Pollan (2001) compares this system to contemporary western production models where agricultural conditions are essentially controlled in order to fit the most desired crop variety. He argues that generally, modern breeding programmes are geared to a western style production system in which the most desired variety is chosen in terms of productivity and quality characteristics, and the growing conditions can be largely controlled. An additional very important characteristic of this type of farming today is the social insurance and subsidy systems which allow farmers to take a bigger risk in cultivation, leading to higher productivity, and providing a backup in case of harvest failures. This is a luxury most farmers in less developed countries do not have, changing the priority from the highest possible productivity to having a resilient system in the first place. As a result, the traditional Andean potato system primarily relied on a high degree of biodiversity to cope with an environment that is very difficult to control.[70]

It should be noted that the traditional Andean potato production system as described here has largely disappeared in many areas in the Andes. Many Andean potato farmers have lost much of their potato biodiversity as market preferences have led to the dominance of only a few varieties. These include both 'cosmopolitan' landraces, as well as improved varieties from potato breeding programs. In large parts of Ecuador, Bolivia and Peru, today farmers mainly cultivate only one or two potato varieties, instead of the sometimes ten or twelve per plot in a traditional potato farming system.[71] A notable exception includes the Peruvian region Huancavelica where still a large on farm biodiversity can be found (CIP and FEDECH 2006).

The existence of such different potato production systems illustrates that breeding and the release of new potato varieties strongly depends on the agricultural production systems that is supported or taken as starting point. The core of the issue lies with the question whether agricultural development in the Andean potato production system inherently requires a replacement of traditional production systems by a more homogeneous industrial model, or whether a non-industrializing approach can be followed. Such an approach might, for example, be characterised by linking formal with informal seed systems, and a combination of elements of a market-oriented production system with the richness in functions of the traditional potato production system.

[70] This does not imply that the ancient Peruvians did not attempt to control or modify their environment. The extensive terracing of the Andean slopes bears witness to the degree to which farmers were in fact able to adapt their natural environment, making it more fit for agriculture.

[71] Personal communication with Graham Thiele.

Challenging a trend towards genetic erosion in potato cultivation in the Andes

Agricultural biodiversity is widely recognized to be a key resource for international food production and plant breeding (Hoisington *et al.* 1999). In fact, the breeding of modern improved varieties essentially depends upon the genetic diversity found in potato landraces, especially when a breeder is looking for the introgression of new traits. To some extent, this has also changed the perspective on agricultural development. From a focus on the introduction of a limited number of improved varieties and a homogenized production system (in terms of genetic diversity and cultivation practices), increasing concern is going out to the replacement of landraces by improved varieties, and the *in situ* conservation of landraces is gaining in importance (Brush 2000).[72] CIP is also aware of the tension between releasing improved potato varieties (which tend to displace traditional landraces), and the importance of conserving crop genetic diversity. Like most CGIAR institutes, CIP has always been active in collecting and conserving the available diversity in potato varieties in its seed bank, which constitutes a first contribution to address the loss of traditional potato varieties. In addition, CIP is involved in more recent initiatives for the *in situ* conservation of potato varieties, like in the Peruvian Potato Park. In this Park, six communities of Andean potato farmers maintain a wide diversity of native potato varieties. The Park is located in an area known as a microcentre of origin and diversity of potatoes, and is an important experiment in the *in situ* conservation of potato diversity, right in its centre of domestication (ANDES and IIED 2005).[73] These initiatives are primarily aimed at conserving what already exists, in a somewhat dynamic (Potato Park) or more static way (seed bank).

More challenging and complex are initiatives to reconsider breeding programmes and potato cultivation in order to counter the trend towards a narrowing down of on farm genetic diversity. Like mentioned, while the traditional Andean potato production system was characterised by a great diversity in traditional potato varieties, today many potato farmers have chosen to specialize on a very limited number of potato varieties for the market. In this context, it is crucial to acknowledge that the decision to shift from the production of traditional potato landraces to modern improved varieties ultimately lies with the potato farmer. A whole range of reasons and factors may play a role in such decisions and a better understanding of the motivations for farmers to cultivate potatoes and to maintain potato diversity is an important precondition to formulating strategies to prevent the loss of this diversity. Some research in this direction is being conducted, and in general, the degree of market integration of Andean potato farmers appears to be an important factor for the specialization on 1 or 2 improved

[72] The concept of 'in-situ conservation' refers to the conservation of traditional varieties through their ongoing cultivation in their natural habitat, in contrast to 'ex-situ conservation'.

[73] See also http://www.parquedelapapa.org/ for more specific information (last accessed 17 September 2008).

varieties that fit the need of urban consumers or the potato processing industry.[74] Having said that, even farmers well integrated in market-oriented production still appear to maintain a significant amount of potato variety on parts of their lands. In fact, a study conducted by Stephen Brush compared two Andean valleys with potato producers with different levels of market integration, and found that farmers in the most 'modernized' valley on average cultivate more different varieties of potatoes, than the farmers in the more traditional valley (Brush 1992). However, it should be noted that these potatoes are grown on small areas and are therefore more vulnerable to random accidents and loss of diversity. While the interests of farmers to keep cultivating a limited set of traditional varieties limits the loss of total genetic diversity, important changes in crop population structure have clearly taken place.

Research is ongoing into the motivations of farmers to maintain this diversity in traditional landraces next to their more homogenized commercial production. Preliminary findings suggest that traditional potato landraces are an important private resource for Andean potato farmers who value the variation in their diet which is dominated by potato consumption. Moreover, potato landraces are used as exchange material and gifts, and are as such associated with indigenous Quechua identity.[75] Thirdly, like already mentioned, on farm agricultural biodiversity has an important function in raising the resilience of the traditional potato production system in high risk cultivation areas. Many of the traditional potato varieties are known for their hardiness in difficult climatic conditions. In a context in which the reliability of a harvest is more important than sheer productivity, the resilience of individual potato varieties, and of the farming system as a whole is of utmost importance. This demonstrates that, next to market value, there is an entirely different set of arguments (cultural, risk avoiding) that farmers take on board in choosing which potatoes to plant. These different motivations for potato cultivation imply that there is a certain scope for the hybridization of traditional and modern production systems.

While it is important to acknowledge that the considerations of farmers which potato varieties to cultivate is not something that can be controlled by CIP, two strategies emerge that can be followed to reduce the genetic narrowing down of cultivated potatoes. First, reconsidering breeding strategies to produce a more diversified output of improved varieties. And second, reconsidering market strategies to challenge the inherent bias to homogenized production. Instead of collecting and conserving all biodiversity in seed banks or potato parks, these options could in fact make the dynamic of potato breeding and cultivation more diverse. However, they are merely theoretical propositions. Is an institute like CIP indeed capable of putting such ideas in practice?

[74] Personal communication with Graham Thiele.

[75] Personal communication with Stef de Haan.

Breeding for diversity

Traditionally, breeding is an affair in which a selection takes place in wide and diverse population of (in this case) potatoes. By crossing different potatoes and making selections, the most suitable potato for a specific production system is developed, selected, and taken further for commercialization. This process inherently means that from the available amount of diversity, only a very limited set of genes is selected for, and will be released for commercial cultivation. Moreover, traditionally a preference has existed for producing a widely popular variety, which has its roots in an incentive structure in which a breeder would be credited for the popularity and wide adoption of 'his/her' variety (Louwaars 2007, p. 41). This means that not only the breeding process itself would narrow down the genetic diversity, but that the marketing of new varieties also has a bias to a limited amount of highly popular and widely adopted varieties. This by the way is by no means specific for potato, but true for commercially cultivated crops in general.

However, this dynamic of breeding programmes and their bias towards a very narrow output can be challenged. There are two activities of CIP that demonstrate that the institute is committed to increasing on-farm biodiversity, and how it can practically contribute to that goal. First, both CIP and INIA (who is responsible for the varietal development on a national level in Peru) are engaged in participatory variety selection trials, that aim to involve farmers in the selection of new potato varieties. In these trials, farmers get to evaluate a number of new potato varieties with potentially new and useful traits. From the e.g. 10 varieties that farmers receive for evaluation, they may choose perhaps 2 or 3 for official variety release. The use of participatory selection or breeding programmes has been discussed in terms of improving seed provision to small-scale farmers (Almekinders *et al.* 2007). However, they may also have a function in widening the output of breeding programmes. For example, non selected varieties that are not officially released, commonly stay in use by the communities involved in evaluation, and after a couple of growing seasons they may prove to have other useful traits which make them more popular. Moreover, it is quite common for such trials to lead to different 'best varieties' that are selected in different communities. In other words, the system explicitly allows for regional and cultural differentiation in preferences of new potato varieties. This is in line with the argument that participatory plant breeding methodologies may encourage the maintenance of more diverse, locally adapted plant populations, to encourage the *in situ* conservation of crop genetic resources, and to lead to the enhancement of genetic diversity (Witcombe *et al.* 2001; Morris and Bellon 2004).

In spite of this generally promising potential of participatory breeding methodologies, an important dilemma lies in the question at what stage to involve farmers. For example, CIP scientists perceive a difficulty in releasing some diversity to farmers, which allows them to actually make a meaningful selection, but preventing the release of too much diversity, which would make it impossible for farmers to efficiently select, meaning that potentially useful material might get lost. In that context it is helpful to refer to a paper by Morris and Bellon (2004) who identify different modes of participatory plant breeding, in which the responsibility of different stages in the breeding process (selection of source germplasm, pre-breeding, cultivar development and varietal evaluation) lie with farmers, scientists, or both. In between the extremes of farmer-led traditional breeding and scientist-led 'scientific breeding', they identify three models: (1) complete participatory breeding in which farmers and scientists are both engaged in all stages of the process; (2) efficient participatory breeding in which farmers are only involved in the selection of source germplasm and the evaluation of varieties; and (3) participatory varietal selection in which farmers are only involved in the most downstream part of varietal selection. While they argue that it is impossible to say which model is most 'optimal' in general, they do raise the point that farmers can have very useful experience and capacities in identifying source material for new breeding programmes, and for varietal selection. Whether farmers can contribute something meaningful to pre-breeding in which mainly material is generated that contains the desired trait, is questionable. Like Morris and Bellon (2004) argue, the question of which model for participation is most appropriate depends on a range of factors, including the characteristics of the crop. For potato, this is extremely relevant since the crop is clonally propagated. This means that farmers are used to dealing with the tubers, rather than with the botanical seed of potatoes. It is therefore difficult to involve farmers in making crosses. Moreover, potato has a complex genetic base, requiring a large amount of crosses in order to develop germplasm with a desired trait into a cultivar that is fit for agricultural cultivation.[76] These factors mean that farmers are likely to have a meaningful contribution in some parts of the breeding process (notably in the identification of interesting parent material and in the selection and evaluation of new varieties), but that interviewed breeders and scientists at CIP but generally don't consider 'complete participatory breeding' to be a feasible or preferred model for potato breeding.

Next to the question of which stages to involve farmers in, a question arises over the traits that farmers can effectively evaluate. While farmers may be very good in evaluating what varieties are useful for them, sometimes a trait like disease resistance may look very appealing during a trial, but may not be very durable in the field. Breeders can make this distinction between vertical resistance (strong but in general not very durable) and horizontal resistance (much more durable). For farmers it is in practice impossible to make such a distinction. A similar problem emerges in breeding for the processing industry, in which a trait such as the sugar content of a potato is crucial. This is also practically impossible for a farmer to evaluate, without the scientific help of a breeder. These examples illustrate how farmer selection is absolutely important, but also has its limitations. Therefore, rather than completely delegating variety selection to farmers, these trials create a domain in which the expertise and science of

[76] Potato is a tetraploid species, making potato genetics much more complicated, and breeding more time consuming than in most other crops that are generally diploid.

formal breeders can be combined with selection by farmers, the combination of which may be expected to work best in really improving agricultural production.

Participatory breeding exercises have to deal with an additional challenge, which is the lack of credibility and statistical foundation of non-standardized farmer trials. This can lead to problems for the formal release of a new and promising variety, and is even reported to undermine the confidence of farmers themselves in participatory selection trials (Morris and Bellon 2004). Some traits that farmers select for have to deal with appearance or culinary value. These traits may be considered to be subjective and are therefore hard to substantiate. Still, for many other traits, a robust but flexible methodology can address this concern. For example, CIP is experimenting with a methodology of 'Mother-Baby' trials, which has been developed by another CGIAR institute (CIMMYT)[77], and in which a centralized trial under reproducible conditions is complemented with a series of satellite trials by farmers (Snapp 1999). The variety of different evaluations from the farmer trials are backed up by statistically sound data from the Mother trial, which allows the formal release of newly selected varieties. Although the methodology can be applied in different ways with different outcomes, this kind of breeding strategy potentially allows for a much wider range of new varieties that will be released, in a much less prescriptive manner, than when a single best-performing variety is released.

Participatory breeding exercises may be an important way of providing farmers with a more diverse set of varieties, adapted to their versatile production system, and encouraging on farm biodiversity. In addition, next to breeding, CIP has been involved in the 'repatriation' of traditional landraces from its genebank to farmer communities who have requested such seed to replace 'lost' local landraces (De Haan and Thiele 2005). This repatriation of potato varieties from the CIP genebank may be an additional way of challenging the loss of traditional varieties and their replacement by a limited number of improved varieties. Concretely, CIP is involved in a number of projects to improve livelihoods of potato farmers in the High Andes. As part of one of these programmes, ALTAGRO[78], CIP has provided a range of potato varieties with different characteristics that were available from the CIP genebank and from previous breeding programmes. This means that no special breeding programme has been set up for this project, but that merely the existing diversity in potato varieties was scanned for potentially useful material. What is significant about this project is that a relatively wide range of material has been released. A CIP breeder identified four groups: (1) native based material with the flavour and quality characteristics that were likely to meet local preferences, with

[77] CIMMYT = '*Centro Internacional de Mejoramiento de Maiz Y Trigo*' (International wheat and maize improvement centre).

[78] ALTAGRO is a CIP project for 'Andean Agriculture in the Altiplano'. See http://inrm.cip.cgiar.org/altagro/ for more information about the project (last checked 11 December 2008).

relatively high yields, and substantial resistance against late blight disease[79]; (2) frost tolerant potato varieties from a previous breeding programme; (3) early maturing varieties which are not particularly resistant to late blight, but because of their early maturing characteristic can escape major stresses like drought and frost; and (4) native potato varieties with relatively high yields and favourable processing characteristics, making them suitable as input for the processing industry. From every group of varieties, some 20 clones were released to farmers for participatory varietal selection. Probably none of the potato varieties mentioned present the perfect potato; if resistance characteristics are favourable, the quality or taste of the potato may be less appreciated. However, this approach of releasing a wide range of potatoes with different characteristics does support a production system in which something is likely to survive in difficult circumstances (like extreme frost), and in which quality characteristics for local consumption are complemented by quality characteristics for a distant processing industry.

All in all, the release of a wider range of potato varieties – either through participatory breeding, or repatriation schemes – means that formal breeding programmes can potentially better cater for a potato production system in which the diversity in varieties is essential to match the diversity in climatic circumstances and soil conditions. This strategy implies a significant departure from a homogenization of production through the release of a limited number of widely performing improved varieties. Clearly, a tension remains in such strategies. The point of breeding is to select for useful potato varieties that can have an added value for potato farmers. The point of a release and certification system is that officially released potato varieties are of an controlled and tested quality, which farmers can safely invest in. Releasing just any set of potato varieties for the sake of encouraging on farm diversity would miss the crucial function of breeding and certification for farmers. Therefore, it remains important to critically evaluate what potato varieties, with what kind of new traits constitute an added value for farmers. The crucial advantage of participatory varietal selection trials, and diversity in repatriation programmes is that farmers themselves can have an important say in what represents 'added value' for them.

Adding market value to diversity

For Andean potato communities, market oriented potato production is crucial for the generation of monetary income. However, markets preferences have generally led to a focus on the cultivation of a very limited number of potato varieties. Next to the fresh consumption of potatoes, an important market has emerged for processed potatoes: chips and crisps. This production process is largely in the hands of a number of multinational companies like Frito Lay and McDonalds, which have strict requirements regarding the potatoes that can serve as input for their production process. Uniformity of potato input is a key factor, in terms of

[79] Late blight is also known as *Phythophtera infestans.*

size and especially sugar content, which is crucial for the process of frying potatoes without browning. This context does not only introduce a bias for the production of a very limited number of potato varieties. In addition, in this context of globalization and market integration, small farmers are often at a disadvantage relative to larger commercial farmers who have superior access to information, services and capital, and who can offer larger volumes of quality products to market agents (Johnson and Berdegué 2004).

This bias towards a homogenized potato production mainly by large scale potato farmers is challenged by the 'Papa Andina' ('Andean potato') project, which is being coordinated by CIP.[80] The project has been set up as an attempt to find new market opportunities for small scale potato farmers in the Andes (Thiele and Devaux 2002; Devaux *et al.* 2007). The project takes as starting point that some commercial opportunities are most beneficial for large scale farmers with a homogeneous production, while other market niches can be served by small scale farmers that have access to a wide diversity in potato varieties. For example, some market niches require small tubers, which in turn require high planting density and manual harvesting. Such constraints in fact provide a competitive advantage to small scale farmers (Thiele and Devaux 2002).

More concretely, a project has been setup to link the production of a mix of native potatoes to marketing in supermarkets in cities. The colourful and tasty native potato varieties turn out to be very attractive as gourmet food, and meet a previously unmet demand. As simple as it is, this T'ikipapa project has been very successful and has won the World Challenge Award 2007, which is awarded to successful business proposals which benefit local communities.[81] The same approach is now also applied to the production and marketing of potato crisps from native potato varieties. Because of the colourful appearance of these 'native potato crisps' they are an innovative and exclusive product. While the scope of this initiative is still limited, the attempt to attach market value to at least part of the existing potato diversity in the high Andes is a very interesting and potentially valuable complementary initiative to the breeding work done at CIP.

The project distinguishes between different types of innovations that may contribute to the creation of new market opportunities: (1) commercial innovations, involving the development of new products or services for market niches to add value to potato production; (2) technical innovations, involving improvements in the way commodities are produced or transformed; and (3) institutional innovations, related to changes in attitudes, habits, or relationships among stakeholders, in order to create more favourable conditions for pro-poor innovation (Devaux *et*

---

[80] See http://papandina.cip.cgiar.org/ (last accessed 17 September 2008).

[81] See http://www.theworldchallenge.co.uk/html/project07_potato.html (last accessed 17 September 2008). Similar approaches exist elsewhere, like in Bolivia where Whipala potatoes are marketed as a mixture of indigenous landraces. See Puente-Rodríguez (2008).

al. 2007). For the initiatives mentioned in this section, it is mostly commercial and institutional innovation that underlies their success. However, the three types of innovation are closely related and the discovery of new market opportunities through commercial innovation can lead to new demands for technical innovation. If potato crisps of native potato varieties are indeed an interesting market niche for Andean potato farmers, this may lead to new breeding and selection rounds on these potato varieties to improve their productivity or processing quality.

Implications: taking local farming systems as starting point

These initiatives of CIP to broaden the genetic base under newly released potato varieties, and to find market niches for the cultivation of traditional landraces reflect the institute's commitment to conserving and valorising the traditional Andean potato production system. With respect to CIP's breeding strategy, it is important to note that it explicitly grants farmers a prominent role in the selection of new varieties and in defining what constitutes added value in new potato varieties. This implies that rather than following an industrializing approach to agricultural development, the institute is adopting an approach that explicitly takes the diversity of local farming systems as starting point, and provides outputs that can be further selected depending on local needs, priorities and preferences. As has become clear, this approach to agricultural development is primarily reflected in the breeding strategy of CIP, rather than in a radical new technological design of the varieties that are being released. But that does not mean that there is no material basis for this alternative strategy to industrialisation. However, this material basis is not captured in the precise characteristics of a single potato variety, but in the fact that the output is no longer a single variety suited for industrial production, but a *collection* of potato varieties, which leaves their evaluation and further selection up to farmers.

In addition, the Papa Andina project clearly demonstrates how important a market-oriented production is in the efforts to improve the economic position of small scale farmers. The discussion of agricultural modernisation, characterised by mechanisms of appropriationism and substitutionism in Chapter 3 has highlighted the risks of linking up with international markets and becoming dependent upon a distant food processing industry that may substitute one agricultural input for another. However, interestingly, the Papa Andina project strongly focuses on the inherent value of native potato varieties of the Andes, both for domestic as well as for international markets. While native potato varieties were traditionally known as a food product for poor Andean farmers, they are now becoming known as wholesome, nutritious food grown naturally in the Andes and an important aspect of Peru's cultural heritage (Devaux *et al.* 2007). This means that far from being an interchangeable industrial input, these potatoes and the way they are marketed emphasize the connection between product and producer. In addition, the project is oriented at finding comparative advantage for small scale farmers who have access to native potato varieties and have the ideal production systems for such varieties. In other words, while the Papa Andina project is definitely about connecting small scale farmers to markets, at least on a conceptual level it challenges trends of substitutionism (interchangeability), and appropriationism (externalization and homogenization of production).

Enabling farmers' seed potato production

Seed is a vital input for agricultural production, but it is not a static input. Seed is alive, it changes under the influence of crosses and selection pressures, it can be diffused, traded and exchanged, and it can deteriorate under the influence of diseases. Seed is supplied by seed systems, which are managing this dynamic of renewal, diffusion and exchange. Graham Thiele (1999, p. 92) defines a seed system as *"an interrelated set of components including breeding, management, replacement and distribution of seed"*. A broad distinction is commonly made between formal and informal (or farmers') seed systems. In the formal seed system the various components of the system such as seed production and supply are operated by public and private sector specialists, and are generally regulated by the public sector. This usually includes an inspection process known as 'certification' and controls over variety release, to ensure that available seed is of a recognized variety with a low incidence of disease. Formal seed systems are generally organised at the (inter)national level, and involve cash transactions and large uniform quantities of seed. In the informal seed system components of local seed selection, seed production and diffusion are managed by farmers themselves, without public sector regulation. They mainly operate at farmer and community level in terms of production and exchange mechanisms. The element of trust which is created by certification of seed in the formal seed system is here created by social networks, and personal ties with reputable traders or neighbours (Thiele 1999; Louwaars 2007).

The importance of formal and informal seed systems differs greatly between different crops, farms, regions and countries. Generally, in developed countries the formal system has to a large extent replaced the informal seed system, especially in the cultivation of potato. In countries like Canada and the Netherlands, more than 90% of the area of potato is sown with certified seed (Young 1990). However, Louwaars (2007) mentions that for the major cereals, the currently estimated use of farm-saved seed in Europe is still 50%, demonstrating that the informal seed system is still of significant importance, even in Europe. In the majority of developing countries, the informal seed system covers 95% of the demand for potato seed (Horton 1987). Turner has demonstrated that in India, the formal system accounted for less than 5% of the total seed use in most major crops, with a maximal percentage of 10% for rice (Turner 1994). Louwaars argues that this is exemplary for developing countries in general (Louwaars 2007). However, the low importance of the formal seed sector does not imply that the formal sector does not have a significant impact on farming. Farmers are not only using traditional varieties, but also diffuse and cultivate scientifically bred varieties, through efficient farmer-to-farmer seed exchange mechanisms (Louwaars 2007).

Potato seed systems

Potato seed systems are somewhat different to most seed systems, because potato is a clonally propagated crop: instead of using seed for its reproduction, farmers replant potato tubers as 'seed potatoes'. A major disadvantage of the use of tubers as 'seed' is that they are relatively bulky and are therefore difficult or expensive to transport in large quantities. Moreover, tubers from a virus-infected potato plant also contain the virus, and give rise to a new generation of potato plants that is already infected. While potato tubers are genetically identical to the parent plants, and therefore in principle have the same agronomic characteristics, the infection by viruses can build up over several generations. This leads to the 'degeneration' of potatoes, which are still genetically the same, but perform poorly because of viral infections. The use of 'botanical seed' of potatoes would reduce this degeneration, since seed does not pass on viruses to the next generation. Moreover, it would make the transport and trading of potato seed easier. The use of potato seed is biologically possible, and CIP has in fact experimented with its use. However, using tubers as seed potatoes increases the chance of germination and provides new seedlings with a portion of energy which increases their survival rate. For that reason, tubers as seed potatoes generally remain preferred over the use of botanical seed.

Considering the importance of tubers as seed potatoes, and the importance to prevent the degeneration of potatoes through viruses, the virus-free multiplication of seed potatoes is a crucial function of any potato seed system. But an important question emerges whether this virus-free multiplication can be part of an informal seed system in which farmers themselves take care of potato seed production, or whether this is a task that is best externalized in a formal potato seed system.

In the Andean region, traditional informal potato seed systems have been known to supply good quality seed potatoes, thanks to the high altitudes at which the degeneration of potatoes is relatively slow. At high altitudes, temperatures are lower and therefore few aphids are found, which are the main vectors for the most damaging viruses (Thiele 1999). This advantage of a reduced virus pressure at high altitudes is used by farmers in lower regions to replenish their 'tired' seeds after a few years of cultivation with new, disease free seed from potato specialists in nearby uplands (Zimmerer 1991; De Haan and Thiele 2005). However, this system was highly typical for the Andean region, and does not have an equivalent in Africa or Asia where potato production is strongly affected by virus infestation. For this reason, in these regions attempts to improve potato production have commonly focused on the supply of better quality potato seed through the setup of a formal potato seed system (Gisselquist *et al.* 1998).

Such a supply of virus-free potatoes is a valuable input for many potato farmers. However, at the same time, the availability of on farm produced potatoes that constitute a 'free' resource of seed potatoes makes monetary investments in formally produced seed potatoes less attractive, and risky.[82] As a result the externalization of potato seed supply as part of its professionalization and formalization has not always been very successful, and not for all farmers.[83] While commercially available virus-free seed potatoes may be an important advantage for some farmers, for other farmers that have to cope with a high risk in cultivation and with little monetary capital to invest, the externalization of seed potato production may simply not be the best way to go. Against that background the question arises of how to contribute to the virus-free production of seed potatoes, and whether alternatives exist to its externalization.

Virus resistance challenging the externalization of seed potato production

Recently, CIP has started to explore the possibility to empower farmers in the production of virus-free seed potatoes as part of informal seed systems, instead of setting up external systems to seed potato production. The key elements of such a strategy would consist of the production of virus resistant potato varieties, cheap and easy diagnostic kits for viruses, and 'positive selection methodologies' for seed potato selection. Virus resistant potatoes would significantly reduce the degeneration of potatoes that are saved from the harvest, and would prevent the build up of virus infections over the years.[84] The use of diagnostic kits would allow farmers to measure the actual level of virus infection in their field, allowing a much more educated decision as to invest in commercially produced (virus free) seed, or not. Moreover, it would allow some sort of quality control of potatoes which may replace the current centrally organised quality system of formal seed producers. Thirdly, the 'positive selection' practice implies that healthy looking potato plants are selected before harvest. The tubers from these plants can then be used as seed potatoes for the next growing season. This contrasts with the common practice of selecting the smallest tubers after the harvest, without consideration of which plants were in fact visibly infected by a virus. In practice, since virus infection tends to reduce tuber size, selection for the smallest tubers implies a selection for virus infection as well. This can largely be prevented by a positive selection practice as described.

The interesting aspect of this approach is that the production of virus resistant potatoes constitutes an example where biotechnology in breeding has an entirely new structuring role in the social relations of production. Where biotechnology has been argued to be instrumental in the industrial appropriation of breeding and multiplication of seed management, in this case it can be instrumental in bringing it back into the domain of farmers' management. This does not mean that the occasional injection of quality seed from an external source would not

[82] See also Snapp *et al.* (2003) for a more general argument about the riskiness of cash investments in cultivation, if market prices of produce can be fluctuating.

[83] Interview data.

[84] In this context, it is vital that potatoes are virus resistant, not just tolerant. While a tolerant potato may still produce in spite of infection, the marketing and transport of infected potatoes might in practice spread virus diseases instead of containing them.

still be a very good idea. However, it would provide farmers with an alternative source of good quality seed potatoes and a more educated decision of when to replenish their potato stock.

Having said that, the material basis underlying this approach to seed multiplication is not all that radically different from other projects, especially with respect to the virus resistance in potatoes. Virus resistance can in fact be a valuable trait in an industrial production system as well, since it reduces the incidence of viral diseases and hence can improve productivity. In other words, virus resistance itself cannot be argued to reflect a non-industrializing approach to agriculture as such. However, what makes the trait special in this example is its embedding within a wider approach to empower farmers in their own seed multiplication, and the combination with diagnostic tests and an improved selection procedure. This leads to the conclusion that an alternative strategy for agricultural development may be reflected in different breeding strategies and different genetic technologies, but that it is the combination of a specific technical material design with its context of application that provides the technology with a structuring role on the production system.

As a final note, it is to be expected that – apart from the technical challenges of actually producing potatoes with sufficient and appropriate virus resistance – this strategy of empowering informal potato seed systems will run up against some serious institutional and regulatory challenges. One of the components of agricultural policy in many countries is the certified production of high-quality seed. Certification of formal varieties provides a mechanism of trust within the seed system, because the quality of the seed is guaranteed. While intended as a measure to protect farmers from bad-quality seed, it also strongly prescribes the mode of agricultural production, since it basically prevents farmers from exchange of their own uncertified seeds.[85] As such, it introduces a strong bias for the external, industrialized production of a limited number of improved varieties, the quality of which can be easily tested, monitored and guaranteed. While this model in potato production has been problematic, the transformation from a certified seed system, to a system that grants farmers the possibility to produce their own, but uncertified seed, may be difficult. In informal seed systems in the Andes, it is a social system that creates trust and so-called 'neighbourhood certificates' for seed sellers. A similar mechanism to address the 'trust component' in seed systems needs to be addressed in some way. The development of intermediate certificates of e.g. 'informal seed, but of tested quality' is suggested by CIP scientists as a way forward, but at this point it still remains a future challenge for the institute to implement such intermediate certification procedures.

[85] See for example recent discussions on the Indian Seed Law (Madhavan and Sanyal 2006). See also the special issue of *Seedling* on seed laws in 2005: (GRAIN 2005)

Working with cultural connotations – Genetic fingerprints as Kipu diagram

The efforts of CIP to reduce the loss of agricultural genetic diversity in the Andes, and to reconsider the externalization of seed potato production, reflect a willingness and ability to take existing production and innovation systems as starting point for agricultural development, and to adopt a predominantly non-industrializing approach to agricultural development. Within that approach the appropriateness of technology development is achieved by broadening the outputs of breeding programmes, by marketing the diversity and quality of native potato varieties, and by empowering farmers in their seed potato production. However, next to these examples of making technology development appropriate within a specific approach to agricultural development, there is an additional dimension of appropriateness. The previous examples – also in the previous chapter on the CIMBAA consortium – have primarily focused on the extent to which farmers are involved in the development process, and on the technical function of new technologies within a specific production system. However, the success of new technologies is also determined by their perception by farmers, and hence by any cultural connotations they may have.

In a negative sense, this is illustrated by the reluctance of CIP to release transgenic potato varieties, even though they are considered to provide a useful means of addressing insect infestation, at least from a technical perspective. Although concerns over outcrossing with other varieties are definitely an additional reason to refrain from the use of transgenic potatoes, it is the public perception of this technology that makes its use definitely inappropriate.

On the other hand, the perception of new technologies and their connotations can be influenced to some extent, embedding the technology in a more favourable context. This approach is illustrated by the publication of a catalogue of native potato varieties in the Huancavelica region in Peru (CIP and FEDECH 2006). For this catalogue, a number of classification methods have been used. Next to phenotypical and taxonomical classification, native potatoes have also been 'fingerprinted' using a number of SSR markers.[86] The use of molecular fingerprints provides an additional 'genetic perspective' on the population of potatoes that has been mapped. Since genetic fingerprints are more precise than the cataloguing of potatoes by their visual characteristics, this technology prevents duplications in the catalogue, or in genebanks. Moreover, genetic fingerprinting allows for an analysis of taxonomic relations and therefore provides an additional perspective on the genetic diversity that has been found, and subsets of closely related potato varieties.

[86] SSR = single sequence repeats; also known as microsatellites. These are DNA markers that allow the tracing and identification of specific individuals by their genetic composition.

In order to present the fingerprinting data in a visually appealing and original way, a system has been developed to represent molecular fingerprints in Kipu-like diagrams (Figure 5.1). The Kipu is an ancient, pre-Columbian system of ropes with knots to represent or register information regarding harvests, exchanges, taxes, *et cetera*.[87] As such, the Kipu bears strong connotations of traditional Andean culture, and admiration for the advanced capabilities of pre-Columbian societies. The representation of SSR markers in the shape of a Kipu diagram is an interesting and creative strategy to present the fingerprinting technology as something that can be easily incorporated into the vision of native potatoes as indigenous material. While it probably does not make the fingerprinting data itself accessible or readable by members of native communities in Huancavelica, it is a symbolic effort to bridge the culturally different worlds of scientific discovery and analysis, and the indigenous communities in which native potato varieties are strongly connected with Quechua identity. As such, this effort is a direct example of the argument made by Brian Pfaffenberger (as discussed in Chapter 3) that it is *discourse* that privileges and legitimizes a specific interpretation of technologies, thereby constituting a political effect (Pfaffenberger 1992).

Apart from the fact that it is an original and charming initiative, this effort to present results of molecular biology in tune with Andean traditions signals an additional dimension in which 'appropriateness' of genetic technologies for agricultural development can be considered. As mentioned before, native potato varieties are associated with Quechua identity, and for that reason their replacement by modern varieties is a painful process that also touches upon

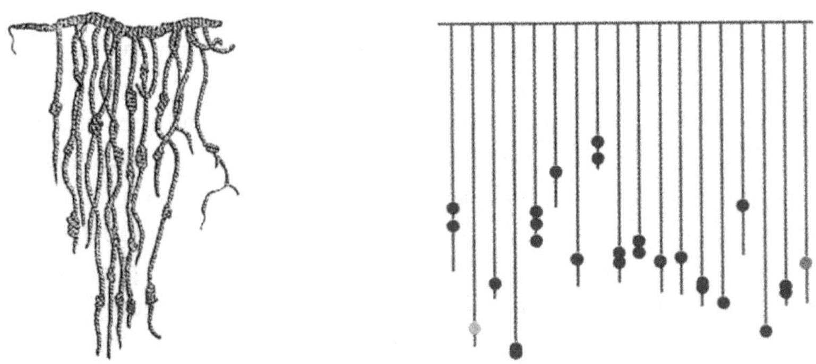

Figure 5.1. A pre-Columbian Kipu (left) and a Kipu-like molecular fingerprint diagram (right). Reprinted with permission from the "Catálogo de variedades de papa nativa de Huancavelica – Perú" (CIP and FEDECH 2006, p. 49).

[87] Also spelled as 'Quipu', see Jacobsen (1964). See also http://en.wikipedia.org/wiki/Quipu (last accessed 17 September 2008).

questions of cultural erosion, next to genetic erosion. The implicit argument in the Kipu-diagrams is that the introduction of new technologies need not be a destruction of traditional cultures, but that a hybridization of cultures is possible. They reflect a concern from the side of CIP to produce outputs that are not only useful in terms of their technical functioning, but also acceptable in terms of their social meaning and cultural connotations.

Discussion – Breeding technologies for a non-industrializing development strategy

The case study of the International Potato Centre has provided an insight in the ways in which the institute aims to provide appropriate technologies and new potato varieties for resource poor farmers. Interestingly, this operationalization moves well beyond instrumental considerations alone, and explicitly relates to the social relations in agricultural innovation. Concretely, the institute acknowledges that it is important to research local priorities for plant breeding and to provide new potato varieties that best address the problems that farmers may face. This can be described as the instrumental dimension of providing appropriate technologies, primarily oriented at the technical functioning of new varieties. However, the different initiatives discussed illustrate that the institute is also more profoundly reconsidering the trajectory of agricultural development. For example, the discussion of how CIP tries to involve farmers in varietal selection, and is experimenting with the release of a wider set of potato varieties rather than a single widely best-performing variety, indicates that appropriateness also means: something adapted to diverse farming conditions and not harming existing agricultural biodiversity. This materializes not so much in a specific new potato variety, but in an approach to breeding and repatriation of potato varieties that is aimed at producing a wider output of varieties, rather than a very limited number of varieties intended for widespread adoption. This approach was complemented by the marketing of native potatoes in the Papa Andina project, valorising the diversity and specific characteristics of the highland potatoes.

What remains uncertain at this point is what the impact will be of this approach on the actual replacement of traditional varieties by modern varieties. In fact, in spite of the interesting efforts to involve farmers in participatory breeding exercises, and to increase the diversity in output, the attractiveness of a homogeneous market-oriented production in order to generate extra income remains big. Moreover, as breeding at CIP gets better and is increasingly able to address traits such as drought and frost tolerance, it is likely that improved varieties will increasingly compete with landraces that currently still play an important role in the cultivation of potatoes in the highest and riskiest parts of the Andes. This means that efforts like Papa Andina for the valorisation of potato diversity, and initiatives for the in situ conservation of traditional landraces, will remain very important and relevant, next to more diversified breeding approaches.

However, like in the other case studies presented in this thesis, this research does not (and cannot) evaluate these initiatives in terms of their technical functioning at this point. In other words, it does not evaluate the discussed initiatives in terms of their effectiveness to actually increase the genetic diversity of improved potato varieties, or to make the cultivation of traditional landraces more attractive. What cán be concluded is that the institute adopts an approach to agricultural development in which the existence of traditional production systems and crop genetic diversity is the starting point for improvements. Breeders at CIP are considering how to add on to the existing potato diversity, rather than to replace it. That in turn reflects an approach to agricultural development that is not relying upon industrialisation, but focuses on peasant based production systems. Moreover, even though much of the research going on in CIP is not participatory at all, it is clear that CIP is modestly challenging the externalization of variety development, by involving farmers in the evaluation and selection of new potato varieties.

A similar challenging of externalization is underlying the attempt to develop virus-resistance kits. While virus resistance has a clear advantage in terms of reducing the incidence of virus infection, this technology also has a specific repercussion in terms of how the potato seed system can be organised. The material design and technical functioning of virus resistant potatoes within the specific context of current potato seed systems has the potential to emancipate farmers in their choice for externally produced seed potatoes, or for their own seed potatoes. While this is unlikely to completely eliminate the need for specialized commercial seed potato production, it does provide farmers with an additional option, and a more educated decision for when to invest in commercial seed potatoes.

The specific example of virus resistance also illustrates that it is difficult and inappropriate to make all too definite statements regarding the importance of the materiality of technological artefacts, in relation to their social meaning within a specific context. The use of virus resistant potatoes in order to empower farmers in their on farm seed potato production is one example in which the technical function of virus resistance has a profound social meaning for the production system. However, this meaning only comes to life in the context of a combined use with virus diagnostic kits and improved selection procedures, and in a socio-economic context in which the commercial prices of seed potatoes are a significant hurdle for potato farmers to replenish their seed stock. The use of virus resistant potatoes in such a project may require no specific technical modification of the potato varieties, but merely an innovative use and embedding of an already existing technology.

Similarly, the material basis underlying CIP's breeding strategies is relevant, but only in a specific context. More than breeding entirely different potatoes, it is an innovative breeding strategy that most clearly reflects a strategy of agricultural development that aims to build on crop genetic diversity rather than to replace it. The material basis of this strategy is not captured in the precise characteristics of a single potato variety, but in the fact that the output is a collection of potato varieties rather than a single variety suited for industrial production. This different material basis in turn explicitly invites evaluation and further selection of varieties by farmers.

In conclusion

All in all, the approach of CIP in aiming to make its breeding outputs and technologies appropriate for resource poor farmers is interesting because it does not only reflect an ambition to provide useful technologies for resource poor farmers, but also reflects an approach to agricultural development that is markedly different from an industrialisation of agriculture. In contrast to the externalization of breeding and seed multiplication as described in Chapter 3, the various initiatives mentioned are characteristic for their attempts to involve farmers in the innovation process. Quite clearly, farmers are not only recipients of new potato varieties, but are also treated as co-innovators in the development and selection of new potato varieties. Moreover, by focusing on the conservation of agricultural biodiversity through a broadening of the breeding output, and the marketing of native potato varieties, the institute treats this biodiversity as a living resource for peasant-based production systems in the Andes, rather than merely as a static input for future breeding programmes.

In conclusion, this case study has illustrated that – within reasonable boundaries – the use of advanced breeding technologies does not necessarily require a further externalization of breeding approaches or seed multiplication. Instead, breeding technologies can be a meaningful element in a non-industrializing approach to agricultural development, provided that not only their technical functioning, but also their structuring role on the social relations of innovation and production are carefully reconsidered.

Chapter 6

Linking upstream genomics research with downstream development objectives – The challenge of the Generation Challenge Programme

> *"The idea of the Green Revolution that we could produce 'Super-variety X' is gone. [...] The idea now is that farmers cultivate certain varieties with a reason, because they are adapted to their needs and preferences. But by introducing a resistance gene or a QTL[88] for drought resistance, we can prevent that every ten years their complete harvest fails because of drought."*
>
> (Anonymous project leader Generation Challenge Programme, October 2007)

Introduction [89]

This chapter shifts the focus from research programmes with clear downstream components (as studied in the previous two chapters), to a research programme that is doing a lot of its research at a much more upstream level.[90] This different position in the innovation process raises entirely different questions regarding the strategy for agricultural development that is supported, as well as regarding the ways in which technology development can be made 'appropriate' for resource poor farmers. While the previous two chapters focused on the social meaning of technological artefacts and breeding strategies in a specific context of application,

---

[88] QTL = Quantitative Trait Locus. This term is used to identify a stretch of DNA that contains or is closely linked with one or multiple genes that are responsible for the expression of a certain trait, in interaction with the environment. The term 'quantitative' signifies that the genetic element is only partly contributing to the expression of the trait, the expression of which also depends on other genetic or environmental factors.

[89] This chapter is largely based upon material previously presented at the CSG/CESAgen conference "Genomics and Society: Setting the agenda", 17-18 April 2008, Amsterdam, the Netherlands. The accompanying paper is published as Vroom(2009): "Linking upstream genomics research with downstream development objectives - The challenge of the Generation Challenge Programme", Tailoring Biotechnologies **4**(3).

[90] The terms 'upstream' and 'downstream' refer to relative positions in an innovation process, with 'upstream' referring to more basic, curiosity driven research, and 'downstream' to applied, problem driven research.

much of the research described in this chapter is too upstream to be analyzed in that way. By consequence, while the previous two chapters focused on the potential for a reconstruction of gene constructs or breeding strategies in a specific context of application, in this chapter the focus shifts to the linkages between upstream research, and downstream applications.

One of the questions that emerges forcefully in this chapter is whether a research programme with such strong upstream components can meaningfully contribute to an agricultural development that would take existing production systems and the innovative potential of farmers as a starting point. As discussed in Chapter 3, concerns over the appropriateness of biotechnologies for resource poor farmers have frequently led to arguments for a locally contextualized, bottom-up technology development. Similarly, the discussion on external and imposed modernisation strategies in Chapter 3 expressed a concern that agro-technology development may reflect and impose a homogeneous and industrial model of agricultural development that is hardly relevant for many resource poor farmers that are producing in marginalized and difficult areas. Hence, the question that arises is whether appropriate biotechnology development essentially requires a local and bottom-up innovation process – implicitly disqualifying any science-led upstream research programme for the resource poor – or that upstream research can in fact provide a meaningful and 'appropriate' contribution to local innovation processes, without prescribing an externally formulated model of agricultural development.

Concretely, this chapter will discuss the efforts of the Generation Challenge Programme (GCP) of the Consultative Group on International Agricultural Development (CGIAR).[91] It will do so by specifically looking at the ways in which this programme aims to connect upstream genomics research to its downstream development objectives of helping resource poor farmers in drought-prone areas. The work of the Generation Challenge Programme is focused on comparative genomics research, and the understanding of genetic mechanisms for drought tolerance in crops. As such, most activities of the programme are rather upstream and relatively far away from the development of concrete new crop varieties. This is in contrast with the CIMBAA project (Chapter 4) which involved upstream genetic research, but was mainly committed to the downstream development of a concrete cabbage variety. Also the projects at CIP discussed in Chapter 5 mostly focused on relatively downstream breeding and selection work. The more upstream character of the Generation Challenge Programme, in combination with the purely public nature of its funding, leads to a research programme with entirely different tensions and dynamics, compared to these other two case studies.

[91] A previous version of this chapter has been presented as a paper titled *"Linking genomics to development objectives – The challenge of the Generation Challenge Programme"* at the CSG/CESAgen conference "Genomics and Society: Setting the agenda", 17-18 April 2008, Amsterdam, the Netherlands.

This chapter will explore the efforts of the GCP to transform its upstream genomics research into actual innovations in farmers fields. For this reason, the first half of the chapter will introduce and discuss the setup of the programme itself and its research agenda. From this discussion, two elements emerge that are crucial in the programme's ambition to link upstream genomics research with downstream agricultural development. The first is the setting of priorities by the programme on an upstream level, in order to develop technologies that will actually address problems that farmers encounter. The second element is the building of an innovation chain that connects upstream genomics research with downstream variety development and testing. While that is an interesting challenge in itself, this chapter aims to take the discussion beyond the notion that strong institutional linkages are crucial in an effective innovation chain. For that reason, the second half of the chapter will adopt an innovation systems perspective and will explore the potential for a more dynamic perspective of how upstream genomics research and downstream innovation systems can be complementary, and what kind of technologies allow for their linkage. As will be shown, this results in an interesting novel function of genetic technologies when they are embedded in a service-like approach, rather than presented as solutions to agricultural production problems as such. Finally, this discussion of the complementarity of different innovation systems relates the case study to the concerns of a homogeneous and imposed nature of agricultural modernisation, as discussed in Chapter 3. Concretely, the chapter reflects upon the extent to which the priority setting by GCP and the innovation system it is constructing, reflect a specific prescriptive strategy for agricultural development, or are flexible enough to serve a multitude of models for agricultural development.

The Generation Challenge Programme – Upstream genomics research for pro-poor agricultural innovation

The Generation Challenge Programme is one of the four 'Challenge Programmes' of the Consultative Group on International Agricultural Research (CGIAR).[92] The CGIAR is a group of 15 international agricultural research centres worldwide that focus on the improvement of genetic resources for agriculture in developing countries. These research centres are diverse in their setup; while some centres focus on a set of the 22 CGIAR mandate crops, others focus on specific agro-climatic regions or policy issues.[93] According to the CGIAR website: *"A CGIAR Challenge Program (CP) is a time-bound, independently-governed program of high-impact research, that targets the CGIAR goals in relation to complex issues of overwhelming global and/ or regional significance, and requires partnerships among a wide range of institutions in order*

[92] See www.generationcp.org for more information about the Generation Challenge Programme and its projects (last accessed 17 September 2008).

[93] See www.cgiar.org for a list of the various CGIAR institutes, their primary focus, and a list of the 22 CGIAR mandate crops (last accessed 17 September 2008).

to deliver its products.'[94] The Challenge Programmes are an institutional novelty within the CGIAR system, in the sense that they are not restricted to a single research centre, but function in between and 'above' the various agricultural research centres. The current four Challenge Programmes focus on 'tapping into crop diversity to improve drought tolerance' (Generation CP), nutritional quality (HarvestPlus), managing food production and water scarcity (Water & Food), and reviving agriculture in Sub-Saharan Africa (Sub-Saharan Africa CP).

The Generation Challenge Programme is committed to the use of comparative genomics, marker assisted breeding, and genotyping technologies to empower plant breeding for resource poor farmers. The underlying rationale is that these kinds of modern genetic technologies are increasingly being used in plant breeding in developed countries, and provide powerful ways of advancing plant breeding, but are difficult to access and use by breeders in developing countries (Ribaut *et al.* 2008). So far, investments in genomic maps of agricultural crops have mainly been limited to a few model crops, or crops of commercial interest to developed countries, while many crops of significance for developing world agriculture have remained 'orphan crops' in terms of research investments (Naylor *et al.* 2005). Moreover, intellectual property restrictions on newly developed technology and biological material often restricts the use of these innovations for agricultural development in 'the South' (Atkinson *et al.* 2003; Louwaars 2007). The CGIAR as a whole is committed to use plant breeding technologies for crops of relevance to developing world agriculture, however its funds are limited and the focus of many CGIAR institutes on specific mandate crops means that potential synergies made possible by comparative genomics have not materialized so far. The Generation Challenge Programme is specifically intended as a cross-cutting initiative to bring together cutting edge genomics research from different institutes, to take advantage of comparative genomics for gene discovery, to build an 'integrated platform of molecular biology and bioinformatics tools', and to facilitate the delivery chain from upstream genomics research to actual innovations in farmers fields (Bruskiewich *et al.* 2006; Generation Challenge Programme 2007).

The GCP is organised in 5 subprogrammes, which focus on different activities, ranging from exploring existing genetic diversity and trait discovery, to developing bioinformatics tools and capacity building (see Box 6.1). Research in each of these subprogrammes is divided in *competitive projects* which have an innovative character, and *commissioned projects* which are more specifically aimed at addressing bottlenecks in the innovation chain, and in making sure that products from earlier research can be validated and taken up into downstream breeding programmes. In addition, GCP makes a distinction between *horizontal* and *vertical* projects, in which the horizontal projects are broad in scope and provide wide platforms of knowledge and methodologies. Making genomic maps of CGIAR mandate crops is such a horizontal activity. The vertical projects in contrast have a much more narrow focus on a specific crop and region, and range from upstream activities down to concrete product development. The

[94] See http://www.cgiar.org/impact/challenge/index.html (last accessed 17 September 2008).

Box 6.1. Subprogrammes of the Generation Challenge Programme.

1. Genetic diversity of global genetic resources.
2. Genomics towards gene discovery.
3. Trait capture for crop improvement.
4. Genetic resources, genomics and crop information systems.
5. Capacity building and enabling delivery.

scope of these projects is limited since they engage with the difficulties and peculiarities of individual farming systems, but their impact can be significant if successful. The hope is that a number of successful examples of vertical projects will provide a proof of concept, a legitimation of the GCP approach, and a starting point for scaling up by other programmes or donors.

Delivery plans, products and users

For the Generation Challenge Programme, one of the main challenges is to make sure that the upstream genomics research actually leads to improved breeding programmes and crop varieties with new traits. This has materialized in a strategy to write a delivery plan for every project that is being funded by GCP (above a funding threshold of $200,000), identifying the concrete 'products' of the research –whether they are genes, markers, germplasm or methodologies- and to identify primary and secondary users of these products. The GCP and its projects are clearly focused on doing upstream research, and therefore cannot engage in the concrete variety development downstream. However, the programme does aim to have an oversight role in making sure that its products cán be taken up downstream, are taken up, and will actually lead to useful innovations for resource poor farmers. In practice, project leaders are required to interact from an early stage onwards with downstream research partners, which are commonly scientists working at national research institutes of developing countries. By involving these people, the outputs of the project are thought to be better tailored to the needs of the downstream partner, increasing chances of a successful innovation process. In addition, next to tailoring research projects to the needs of downstream partners, the building of capacity of these downstream partners is a focal point, in order to ensure that they can actually take up the genetic information or technologies that are being produced by GCP. Both the development of delivery plans, and organizing capacity building of downstream partners are important objectives of subprogramme 5 (SP5).

This focus on delivery and capacity building at national research partners demonstrates how GCP operationalizes its commitment to having an impact. In doing so, the GCP cannot possibly control and organise all steps in the innovation process going from upstream gene discovery to varietal development in a specific context. In spite of that restriction, the facilitating role of the programme and its oversight role are meant to create favourable conditions for the development of actual innovations in terms of new crop varieties for resource poor farmers. This output oriented character of the programme and the explicit focus on identifying products and different levels of users for every project is an important characteristic of GCP. It is typical for contemporary debates on the effectiveness of public sector agricultural research, on 'innovation systems' for agricultural development, and with ideas to stimulate interaction between different institutes and actors in order to bridge the gap between invention and innovation. But, next to making sure that bottlenecks in the innovation chain are being addressed, the question emerges what kind of farmers GCP actually targets, and with what kind of strategy for agricultural development. In other words, we need to shift focus from the *procedure* of delivery, to the *target(s)* and *objectives* of delivery.

Targeting the poor – Farming systems, crops and traits

Like any research programme, the GCP is caught in between ambitious goals and a limited amount of funding and lifetime. This has led the programme to execute a priority setting exercise in order to define a coherent set of principles for the selection of projects to be funded by GCP. Since priority setting exercises force a programme to limit its activities and potential beneficiaries, they expose how formal priorities are operationalized in practice, in the selection of projects. The GCP priority setting exercise and its outcomes are elaborated below, and their impact on targeting resource poor farmers are discussed.

In terms of discussing how the outcomes of GCP can reach resource poor farmers specifically, the focus of the programme on drought tolerance is a first important element. Drought as focus trait has been a leading principle right from the beginning of the programme for a variety of reasons. One is the global importance of drought stress on agricultural production (Moffat 2002; Ribaut 2006; Tuberosa and Salvi 2006), and hence the potential to have a great impact if drought resistance can be successfully managed. Moreover, considering that the GCP would not focus on a single crop, or small set of crops, a trait was chosen that is problematic across a wide range of crops. Thirdly, drought tolerance is a trait that is very difficult to tackle because of it is complexity. In interviews, some scientists jokingly refer to it as the 'holy grail' of plant breeding for the developing world. The supposed value of comparative genomics and high-throughput analyses for pro-poor plant breeding would be best demonstrated by addressing a trait that has been notoriously difficult to crack in the past. Drought tolerance is just such a trait.

In practice, the focus on drought includes a number of drought related traits such as aluminium toxicity, phosphorous uptake and in some cases completely different traits that turn out to be main limiting factors for production. For example, in some cases drought is an important limiting factor for production, but improved drought resistance only provides added value if a certain disease or pest is addressed as well. This can be a reason for specific GCP projects to widen their focus on different traits, especially in the downstream variety development.[95] Moreover, it is important to note that drought as such is a very complex trait, which is manifest through very different mechanisms. In practice this means that research is focused on different genetic mechanisms which are of relevance in different drought prone environments, for example depending on the moment of drought stress early or late in the growing season.

With drought tolerance as priority trait, the priority setting exercise conducted by GCP needed to identify areas where the need for improved food production was highest, where drought was indeed a serious limiting factor, and where there was a scope for improvement through the release of improved seed varieties. Only in that context would the work of GCP have a significant added value. A farming systems approach has been taken as starting point for identifying what reasonable target areas could be. In 2001, John Dixon *et al.* (2001) developed a list of farming systems related to poverty in six main developing regions.[96] Two types of criteria were used to identify these farming systems: first the available natural resource base, climate, typography, farm size and tenure; secondly, household livelihood patterns, technologies and farm management and organisation. These criteria led to a list of 72 distinct farming systems, with an average agricultural population of about 40 million (FAO 2001). Together, they form a global map of agricultural production in the developing world that cuts across national boundaries and provide a novel view on their constraints and potentialities.

GCP has chosen to take these farming systems as heuristic for identifying problem regions, and selected for the farming systems in which chronic poverty was apparent, and drought a major limiting factor for production (Generation Challenge Programme 2006). As poverty indicator, the number of children that are stunted in their growth (poor growth in length for age) had been chosen, since this indicates malnourishment over a prolonged period in time. A number of 2.5 million stunted children per farming system was set as threshold. In order to select for drought stress in a farming system, a failed season drought model was developed, indicating what areas would be most prone to drought stress in agricultural cultivation. The overlaying of these data on the map of global farming systems led to a selection of priority farming systems in which chronic poverty was a major issue and drought a major limiting factor for agricultural production. After leaving out a number of farming systems based on trees or pastoralism (farmers with livestock) for which different mechanisms of poverty are thought to play a role, a set of *15 target farming systems* remained. In order to determine whether crop research could indeed be an appropriate response to the problems in these target

[95] A concrete example is a project on cassava, in which new genetic variation from Colombia is brought to Africa in order to address drought tolerance. However, in validating the material in Nigeria, virus infestation appeared to be a major limiting factor, not only to production, but also to the breeding programme itself. Although the primary focus of the GCP project is on drought tolerance, the same genetic tools and knowledge can be used to start selecting for virus resistance as well.

[96] These six regions are: Sub-Saharan Africa, Middle-East and North Africa, Eastern Europe and Central Asia, South Asia, East Asia and Pacific, Latin America and Caribbean.

farming systems, the productivity of the major crops in these systems was assessed in order to verify that the chronic malnourishment that was identified indeed corresponded to poor agricultural production (Hyman *et al.* 2008). Since this was the case, it was concluded that genetic improvement could potentially help to increase food production and hence alleviate poverty. Next to identifying 15 target farming systems, a number of *13 priority crops* were identified that represent 95% of the cultivated area in those farming systems.[97] See Table 6.1 for an overview of the priority farming systems for the GCP and their main crops.

15 target farming systems, 13 priority crops and one priority trait: that is the main result of the priority setting exercise as conducted by the Generation Challenge Programme. The list of target farming systems and their major crops provides a concrete narrowing down of the research focus for the Generation Challenge Programme. It helps to concentrate the GCP funds on a limited number of crops, and prevents a dilution of funds and a loss of impact of the research. But what do these priorities mean in terms of research agenda, and the addressing of needs of the resource poor?

Challenges for a science-led research programme

The aim of the Generation Challenge Programme is that its upstream genomics research into mechanisms of drought tolerance leads to outputs that are ultimately relevant for resource poor farmers in drought prone areas. The priority setting exercise as elaborated above has provided the programme with a legitimation for focusing on drought, and has indicated in what farming systems and crops this research is most likely to have an impact. In fact, this priority setting exercise is a very explicit way of technically defining what would constitute an appropriate technical output of the GCP. What is remarkable in that respect is that the operationalization of appropriateness has taken place in relative isolation of a concrete farming system, and the problems that farmers encounter in real life. Clearly, there is a wide legitimacy for addressing drought tolerance, but it seems fair to conclude that the research agenda is primarily science-led. This raises the question to what extent such a technical and science-led operationalization of appropriateness is legitimized in order to produce useful technologies for resource poor farmers, and to what extent it represents a specific perspective on how agro-technological development should be organised.

In this respect, a first observation that can be made is that the legitimacy of a top-down science driven approach to some extent depends on the kind of trait the programme is working on. For some traits – or technical solutions in general – a very precise and locally specific understanding of the problem is essential for successfully solving it. Disease or pest resistance is such a trait, which strongly depends on local climatic conditions, crop ecology and local

[97] These 13 crops are: barley, beans, cassava, groundnut, maize, millet, potato, pulses (specifically cowpea and chickpea), rice, sorghum, sweet potato, and wheat (Hyman *et al.* 2008).

Table 6.1. Priority farming systems for the Generation Challenge Programme (Hyman et al. *2008). Regions are listed in order of highest to lowest absolute number of stunted children per farming system. For each farming system, the main crops are given.*

Region	Farming system	Crops in farming system
South Asia	Rice-wheat	Rice, pulses, (chickpea), millet, wheat, maize, bean
South Asia	Rainfed mixed	Rice, millet, sorghum, chickpea, bean, groundnut, maize, wheat
East-Asia and the Pacific	Upland intensive mixed	Maize, rice, wheat, sweet potato, potato, bean
East-Asia and the Pacific	Lowland rice	Rice, maize, wheat, sweet potato, groundnut
South Asia	Rice	Rice, pulses (chickpea)
Sub-Saharan Africa	Cereal-root	Sorghum, millet, pulses (cowpea), maize, groundnut, cassava
Sub-Saharan Africa	Maize mixed	Maize, cassava, sorghum, pulses, groundnut, millet, bean, sweet potato
South Asia	Highland mix	Rice, maize, wheat, potato, groundnut, pulses (chickpea)
Sub-Saharan Africa	Root crop	Maize, cassava, rice, sweet potato, cowpea, sorghum, groundnut, bean
South Asia	Dry rainfed	Sorghum, millet, chickpea, groundnut, bean
Sub-Saharan Africa	Agropastoral millet/ sorghum	Millet, sorghum, pulses, groundnut, maize
Latin- America and the Caribbean	Maize-beans	Maize, bean, sorghum
Sub-Saharan Africa	Highland temperate mix	Maize, wheat, sorghum, barley, millet pulses
East-Asia and the Pacific	Temperate mixed	Maize, wheat, potato, groundnut, millet
East-Asia and the Pacific	Highland extensive mixed	Rice, maize, wheat, potato, groundnut, pulses

strains of viruses. The way this applies to drought tolerance as a trait depends much on the understanding of this trait. As mentioned before, drought tolerance in general is a trait that can be useful for many farmers, regardless of their precise circumstances or cropping system. However, taking a closer look at the trait reveals its complexity, its frequent co-occurrence with other problems with soil fertility or toxicity, and with the vulnerability of crops to pest infestation. This means that in practice, a very local understanding of the production constraints will probably be needed to successfully develop new crop varieties that can deal with the specific circumstances on a higher level of detail.

Having said that, it is possible that a limited number of genetic mechanisms in fact determines drought tolerance across a wide range of crops and circumstances. For example, the ability of a crop to *perceive* drought and to adapt its metabolism accordingly may be regulated by a very specific genetic switch, in a wide range of plants. The discovery of such a mechanism would mean that basic genetic research can have a great influence on the development of drought tolerant crops, in spite of the different manifestations of drought stress/tolerance in different environments. In that sense, the success of a generally science driven approach to address a trait like drought tolerance will depend on the genetic mechanisms that play a role, which in turn means that it is an empirical question which will only become clearer during the research of GCP itself.

However, next to the technical variation in what drought stress means, there may be completely different factors at play that limit agricultural cultivation, and that remain opaque in a merely technical view on what appropriate technology is. For example, while a specific farming system may be characterised as 'drought prone', that does not mean that providing drought tolerant varieties is the most apt way of addressing poor agricultural performance. In fact, actual problems in cultivation may have less to do with the poor quality of the varieties being grown, than with political decisions on land use within a country. Take for example the 'maize-beans farming system', which is the only target farming system of the GCP in the Americas. If this farming system is diagnosed with sub-average production levels, genetic improvement of maize or beans may clearly be one potential solution. However, in many Central American countries, an unequal division of fertile lands is a well known and historical problem since the colonisation of the continent (Morley 2001). If the most fertile lands are in the hand of a small rich elite, and to an important extent dedicated to export crops such as coffee or fruit, the reason that many other farmers have sub-optimal yields may as much be related to the division of fertile land, as to the quality of their germplasm (Garst and Barry 1990). Clearly, such locally diverging, socio-political dimensions of problems in agricultural production are not taken on board in the research priorities of the GCP.

The argument that arises is that a priority setting exercise based upon statistics of chronic undernutrition and models of drought stress on agriculture may provide a legitimation for focusing on drought tolerance in certain crops for certain regions, but that a more localized understanding of the problems and potential solutions may be required to really contribute something useful for agricultural cultivation. This creates some tension between a general focus on upstream genomics research to drought tolerance, and the diversity in local manifestations of the problem at hand. How the Generation Challenge Programme addresses this tension will be discussed in the upcoming sections.

A final challenge that arises is related to actually reaching the resource poor farmers that are targeted by the research programme. Arguing that specific genetic technologies can have an impact in farmers' fields is not the same as arguing that the impact is especially relevant in the context of poverty alleviation. Andy Hall *et al.* concretely describe a concrete manifestation of this problem in a project on the improvement of post-harvest conservation of mangos (Hall *et al.* 2003; Hall *et al.* 2004a). In this project, technical assistance was intended to benefit resource poor mango farmers, but Hall *et al.* report that in practice the innovation process was dominated by large-scale, non poor mango producers who were most actively involved in mango export. By failing to investigate stakeholder agendas at an early stage of the project, not only innovation itself was impeded because of institutional constraints, but – at least as important – the chance that anything coming out of the project would actually benefit resource poor mango farmers was very low.

Obviously, the kind of downstream research partners that GCP works with matters enormously in order to make sure that products do not only reach farmers' fields, but to make sure that they also actually contribute to poverty alleviation among small scale farmers. Concretely, this requires GCP to go further than making sure that downstream partners are able to take up the outputs of upstream genomics research and translate them into new crop varieties. Instead, it requires the programme to very consciously evaluate who the farmers are that are targeted by the downstream research partners, and are therefore most likely to benefit from the GCP products.

Note that the appropriateness of technological development here no longer primarily depends on the material design of the drought resistance technology. In fact, in spite of the claim that drought tolerance as trait is essentially of most relevance for resource poor farmers in drought-prone areas, it might as well as unlock new arid areas for industrialized farming that would not have been remunerative for cultivation without better drought resistant crops. In other words, while in a concrete context of application the precise material design of new crop varieties can be expected to be of relevance, on the more upstream level at which GCP works, it is of crucial importance to embed the technology in an institutional setting that ensures its appropriateness for resource poor farmers, rather than relying on the trait is relevance for resource poor farmers alone.

Complementary innovation systems

In discussing the challenges for GCP to contribute to agricultural development for resource poor farmers, the issues that emerge are in fact directly related to the tension between bottom-up versus top-down innovation trajectories. While a bottom-up technology development project may be closely in touch with local needs and circumstances, it is unlikely to engage in comparative genomics research and to harness its potential to find new genetic mechanisms of dealing with drought stress. It takes a significant amount of upstream scientific work to actually be able to unlock this 'genetic potential' which is thought to be present in existing germplasm collections. On the other hand, a science led activity will always have difficulties in dealing with the peculiarities and complexities of different production systems 'on the ground'. It may provide generally applicable solutions, but their adaptation to different local situations is quite another challenge, which is generally taken up by local research institutes, extension services or development projects. But does that mean that science-led innovation in new crop varieties is inherently less appropriate than bottom up innovation processes?

Rather than continuing the top-down versus bottom-up dichotomy with its apparent contradictions, it seems more appropriate to explore the complementarities between both approaches, like advocated by the Systems of Innovation framework. This conceptual framework has recently been adopted by a number of scholars as a new perspective on the question of how to organise biotechnology development for resource poor farmers. The framework goes back to the conceptualization of 'National Systems of Innovation', as originally developed by authors like Freeman (1987), Nelson (1993) and Lundvall (1992), somewhat more recently reviewed by Charles Edquist (1997), and summarized in relation to development issues by Andy Hall (Hall and Yoganand 2004; Hall *et al.* 2004c; Hall 2005).

According to Hall (2005) the concept of National Systems of Innovation emerged because conventional economic models had limited power to explain innovation, which was viewed conventionally as a linear process driven by research. In contrast, the innovation systems framework sees innovation in a more systemic, interactive and evolutionary way, whereby networks of organisations, together with the institutions and policies that affect their innovative behaviour and performance, bring new products and processes into economic and social use. The framework has become rather popular during the last decade and is frequently used to understand and strengthen innovation at national, regional and sectoral levels.

The conceptual framework of systems of innovation consists of a wide, but somewhat diffuse body of literature with various approaches. In addition, these approaches have changed over the last few decades (Smits and Kuhlman 2004). In Box 6.2, some key points are summarized that characterize contemporary thinking about Systems of Innovation.

The Systems of Innovation framework provides a helpful tool to conceptualize the innovation process and the role of different institutions and activities as part of the larger innovation system. However, it does not provide a single model of how innovation should be organised. For example, Andy Hall has concretely explored the importance of agricultural innovation systems, and has indicated both their diversity and complementarities. Rather than formulating the ideal type agricultural innovation system, he elaborates a genealogy of different types of innovation systems in international agricultural development (Hall 2005). This genealogy ranges from highly science driven public sector research, via R&D led agribusinesses, to pro-poor participatory innovation for complex agro-ecologies. More important than the exact

Box 6.2. Key elements of the Systems of Innovation framework (Hall and Yoganand 2004).

- The Systems of Innovation framework has a focus on innovation processes, rather than on mere production of knowledge.
- The framework conceptualizes research as part of the wider process of innovation, and therefore helps in identifying the scope of the actors involved and the wider set of relationships in which research is embedded.
- It breaks out of the dichotomy between technology-push and demand-pull theories. Instead, it recognizes that both processes are potentially important at different stages in the innovation process.
- It recognizes that the institutional context of the organisations involved promotes dominant interests and shapes the outcomes of the innovation system as a whole. It therefore urges to examine and reveal which agendas are being promoted, and highlights the arena in which the voice of the poor can be promoted.
- It recognizes a system of innovation as a social system, and therefore does not just focus on the degree of connectivity between different elements, but on learning and adaptive processes.
- It is a framework for analysis and planning, and not restricted to a single disciplinary convention.

typology of agricultural innovation systems, he argues that the recognition of diversity in innovation systems is important for a number of reasons:

> *"Firstly, it allows policy and capacity development activities to recognize and support the co-existence of different types of innovation capacity. This helps break out of the false dichotomy whereby old practices are vilified at the expense of new without recognizing synergy. Secondly, it allows emphasis to be given to ways of strengthening the strategic, purpose-oriented interaction of these systems at various points of intersection. This shifts attention to complementing and integrating different ways of producing and using knowledge rather than arguing for homogeny and, for example, insisting that all approaches having to become participatory or partnership based or that all approaches have to be science-led. Clearly neither of these propositions is workable and could undermine well intentioned capacity development efforts."*
>
> (Hall 2005, p. 627-628)

Hall stresses the complementarity of different innovation systems and the roles that different institutes can play. But at the same time, he acknowledges that a challenge still lies ahead in actually linking different types of innovation systems and for example bringing biotechnology innovation systems to bear on farmer participatory innovation systems. This is exactly the challenge for GCP in linking up with different downstream research partners. A potential complementarity may be observed between the upstream research that GCP funds, and the engagement with the complexity of farmers on the ground that farmer participatory innovation systems are best at, and which is more typical for GCP's vertical research projects. But the question is how this complementarity can be best exploited in practice, and whether an innovation chain in which upstream outputs are transformed into a series of downstream inputs is the most useful heuristic. Considering the interest in technology development that takes local innovation capacity and diversity in production systems as starting point (as expressed in Chapter 3), it may be worthwhile to consider how GCP outputs can be valuable in a wider range of downstream development trajectories, and what would constitute an effective technical interface between upstream genetic resources and downstream applications.

Complementarity in practice: different research partners and technology as a service

The first attempt of the Generation Challenge Programme to exploit the complementarity in science-driven and bottom-up innovation processes is by strategically investing in the aforementioned horizontal and vertical research projects. While horizontal research is more upstream and focused on the general understanding of genetic mechanisms behind drought tolerance, vertical research projects are committed to the delivery of concrete new crop varieties, adapted to the specific problems of farmers in a specific region. These vertical research projects are also responding to prioritized needs of farmers, beyond the focus on drought tolerance, and hence they are argued to be more demand driven than horizontal research projects. Because of their more downstream and focused nature, they do have to take into account what the concrete manifestation of drought stress is on a local level, and how problems of agricultural production are often a combination of a wider set of factors. Whether this is a successful approach in practice depends on the evaluation of the specific vertical research projects. However, in terms of linking upstream research with downstream variety development, this approach demonstrates the possibility to combine different innovative dynamics within the same programme, ideally leading to an optimal integration of perspectives and interaction between different levels of innovative activity.

In addition, as becomes clear out of an inventory of its projects, the GCP is capable of linking up with a wide range of downstream research partners, with different technical needs and objectives. The most obvious downstream research partners for the Generation Challenge Programme are the national agricultural research institutes in any developing country. These are the kind of institutes that have been beneficiaries of CGIAR related research for decades, and which generally play a central role in the agricultural policies and research of developing nations. However, in practice, the range of partners appears to be wider than just these national research institutes. A nice case in point is the recently started project on common bean (*Phaseolus vulgaris L*), which is focused on the Maize-Beans target farming system in Central America, and focuses on drought stress and diseases that occur under drought and low soil fertility conditions. The project is headed by a scientist from INIFAP[98], the Mexican national agricultural research institute, but has partners in Mexico, Nicaragua, Cuba and Haiti. What makes this project special is that the partners from the countries involved, are very different in character. While the development of varieties and dissemination of seed in Mexico will be carried out with an organisation of bean producers, the partner in Cuba is the national research institute that will use the GCP outputs for participatory breeding with Cuban farmers. In Haiti, little public investments are made in agricultural research and extension, and a national agricultural research centre is lacking. Therefore, contact has been made with an NGO ('ORE') that will be engaged with the dissemination and evaluation of new bean varieties that will be produced as part of the GCP project.

These completely different downstream organisations demonstrate a certain flexibility of the GCP in terms of linking up with different research partners. This observation is significant against the background of the discussion in Chapter 3 on the prescriptiveness of homogeneous modernisation processes. The variety in downstream research partners and their approaches to agricultural development demonstrate that the upstream GCP research does not preclude a diversity in its operationalization on a downstream level and that it does not put a major restriction on the kind of collaborations that are set up. In addition, the diversity in downstream research partners illustrates that rather than an innovation chain, an innovation network may be a more appropriate heuristic to map and represent the interactions between the upstream GCP innovation system and a range of different downstream innovation systems.

Technology as a service – The Genotyping Support Service

The complementarity between horizontal and vertical research projects, and the collaboration with different types of downstream research partners may indicate an institutional flexibility. However, in practice it does introduce new technical requirements for the outputs of GCP. This is for example visible in the differences between the participatory breeding programme in Cuba, and the NGO in Haiti that does not have the capacities to get involved in breeding, but will focus on dissemination of improved seeds. For the GCP project, this means that outputs will have to be attuned to different needs; concretely that a wider breeding population with some diversity will have to be provided for the participatory breeding exercise in Cuba, while a limited set of finished varieties can be disseminated in Haiti. However, the question of what kind of technical interface would support a meaningful complementarity between science-led and bottom-up innovation has a wider relevance.

[98] INIFAP = *Instituto Nacional de Investigaciones Forestales, Agrícolas y Pecuarias.*

The GCP produces a wide range of 'technical' outputs, which include knowledge about genes, traits, molecular markers to introgress such genes, and potentially parental material containing new traits (like drought resistance) to be incorporated in breeding programmes. In addition, GCP creates communities of scientists working on the same crop, technology platforms and genomic maps of crops that are of wider use and significance. These outputs may be useful in wide range of settings, but they do require a certain level of expertise to be taken up in further downstream research or variety development. The most straightforward way of dealing with this problem is to invest in downstream capacity building, which is the main burden of subprogramme 5. The activities of this subprogramme include the setup of training courses, exchange programmes of scientists, and other capacity building activities that are primarily aimed at increasing the level of up-to-date knowledge about genetics, the interpretation of genetic data and the use of molecular markers in breeding programmes.

In addition, a more structural problem has been identified with the use of molecular markers and other genetic analyses because of the lack of basic infrastructure with many research partners in the south. Access to electricity, clean water, personnel for technical assistance, and reagents may be structurally difficult, and not easy to overcome by a couple of capacity building exercises, or investments from the GCP.[99] For this reason, a Genotyping Support Service (GSS) has been set up that allows the outsourcing of genetic analyses to specialized institutes in developed countries at competitive prices. The GSS plays an intermediate role in helping research partners to develop research proposals, in linking with specialized genotyping institutes, and in interpreting the data that come out of the exercise. While the exact genotyping activity can be tailored to the needs of the research partner, in general two kinds of major categories are distinguished in the proposals for the GSS: (1) determining the genetic diversity in breeding material or germplasm bank accessions, and (2) running marker assisted selection in a population on a trait for which markers are available.

Although the Genotyping Support Service can be seen as a way to overcome bottlenecks in using the other outputs of the GCP (like molecular markers for drought), it also allows for completely different uses, and interesting synergies between different projects. Take the example of groundnut research at three different partners of GCP in Brazil, Bolivia and Senegal. Groundnut (*Arachis hypogaea L.*) is the most widely cultivated legume in Africa, with most of the production originating from drought-prone areas. Drought considerably reduces yield and production. Cultivated groundnut has a narrow genetic basis and the first step for improving drought tolerance in this crop is by enhancing genetic diversity. GCP is involved in a project in Senegal, one of the main groundnut producers in West Africa, aimed at evaluating new sources of drought resistance in groundnut. The Brazilian agricultural research institute

[99] See also similar experiences of the Convergence of Sciences project, as reported by Richards *et al.* (2009).

EMBRAPA[100] had earlier been involved in projects aimed at widening the genetic base of Brazilian groundnut production, and had been successful in making some distant crosses between cultivated varieties and wild varieties of groundnut. Meanwhile, in Bolivia – the centre of origin of groundnut and still home to a wide diversity of the crop – PROINPA[101] is involved in mapping and conserving groundnut diversity. Like in many countries, the diversity in traditional landraces has come under pressure with the introduction of improved varieties from developed countries. PROINPA has a mandate from the Bolivian government to be a curator for certain germplasm collections, and as such is concerned with maintenance of existing crop diversity. In addition, the foundation is interested in breeding with traditional landraces in order to provide Bolivian groundnut farmers with varieties that both have good agricultural characteristics, and fit the needs of Bolivian consumers.

A problem for PROINPA is the collection and categorization of their collection of groundnut germplasm, which makes its valorisation for future breeding activities more difficult. Conventionally, phenotypical (morphological) characteristics were used to categorize accessions of the seed bank. However, since the appearance of groundnuts (and other crops) can be strongly influenced by environmental conditions, this is a rather inefficient and unreliable way of mapping diversity. In order to get a better idea of the real genetic diversity that is present in the Bolivian groundnut collection, a proposal was submitted for the Genotyping Support Service in order to fingerprint the collection, and to map the diversity that is present. This resulted in a set of genetic information that provided a much more precise image of the diversity within the collection, and its position vis-à-vis other groundnut collections. Moreover, it became clear that the Bolivian collection contains a number of accessions that are unknown to other seed banks of groundnut, and are therefore potentially valuable sources of new traits.

For PROINPA in Bolivia, this exercise was about getting a better idea of the diversity that is present, in order to make better use of it in future breeding programmes. However, obvious synergies emerge with the work of EMBRAPA in Brazil and the GCP project on groundnut in Senegal. While EMBRAPA has experience with actually widening the genetic base of groundnut through crosses with wild relatives, the additional genetic diversity that can be created this way may be very valuable for the project in Senegal. What this demonstrates is that the combination of gene discovery, marker development, marker assisted breeding and genotyping technology can be used in very diverse context, with different objectives.

In addition, a fruitful link between the Genotyping Support Service and 'pro-poor participatory innovation for complex agro-ecologies' can be indicated. Already, research partners are not required to be able to extract DNA from their germplasm collection: instead, providing tissue material from plants to be screened is sufficient for the specialized genotyping lab to extract

[100] EMBRAPA = *Empresa Brasileira de Pesquisa Agropecuária.*

[101] PROINPA = *Promoción y Investigación de Productos Andinos.*

DNA and to perform the analysis. If this kind of service is not only provided by GCP but becomes an integrated service of many national research institutes, it is only a small step to offer local and small scale participatory breeding programmes to provide marker assisted selection, *in addition to* the participatory selection with farmers. The value of such a service would lie in the selection for traits that are difficult to assess by farmers, because they are not visible to the eye. Examples include the processing quality of potatoes, which depends on the sugar content of potatoes, or horizontal disease resistance which depends on the presence of a number of quantitative genetic elements. For farmers it is nearly impossible to assess the difference between strong vertical resistance that easily breaks down, and somewhat weaker horizontal resistance that is expected to be much more durable.

If selection on such difficult traits by markers is *complemented* by a participatory breeding programme in which farmers themselves can identify what kind of new varieties would suit their production best, we can speak of a truly successful linkage between the complementary capacities of molecular scientists and farmers. The effects of such an approach would go beyond making breeding more efficient. In terms of social relations of innovations, it implies that genetic approaches to breeding do not require an externalization of the breeding and selection process, but rather that the genetic perspective becomes an integrated part of an innovation process that is essentially led by farmers themselves, as part of a participatory breeding programme.

Discussion – The potential for a service-like approach to agro-technological innovation

This thesis is investigating how genetic technologies for agriculture are made appropriate for resource poor farmers in three very different projects, and how the operationalization of appropriateness relates to a specific strategy for agricultural development. As has become clear from this chapter, the analysis of the appropriateness of technologies, and their relationship with a specific strategy of agricultural development depends significantly on the relative position in an innovation system that is being studied. While the case studies in the previous two chapters allowed for a more or less contextualized understanding of the function of breeding technologies, and their social meaning within a concrete production system, a similar analysis on the level of upstream genomics research makes little sense. The argument is not that at an upstream level no assumptions are being embedded in the development of technologies, or in the selection of priorities. Clearly, also the Generation Challenge Programme makes specific choices regarding the farmers it is targeting and the traits that are of most interest to those farmers, as illustrated by the GCP priority setting exercise.[102] However, what has become clear is that the development of genomic maps, the understanding of genetic mechanisms

[102] In addition, consider the discussion on the assumptions taken on board in the CIMBAA project at an early phase in the innovation process; Chapter 4.

underlying drought tolerance, and especially the setting up of a Genotyping Support Service can become instrumental in a great variety of ways and in a great range of downstream projects. For that reason, it makes little sense to discuss or define the appropriateness of these technologies – or rather scientific results – as such. Instead, the relationship of this upstream domain with more applied technology development constituted a more relevant locus for investigation.

In discussing this relationship between the upstream research and downstream applications, an important argument has been borrowed from Andy Hall (2005) to move beyond a dichotomy between top-down and bottom-up innovation dynamics, and to look at the complementarity of these approaches instead. In exploring this complementarity in the context of the GCP, it becomes apparent that the kind of technologies and knowledge that GCP produces are to an important extent enabling technologies, or tools in further research. Rather than creating concrete artefacts that farmers can use in their production, the production of genomics maps, knowledge about genetic mechanisms, or in a later stage molecular markers allow downstream research partners to advance their breeding programmes on a local basis. This means that a potentially very important distinction emerges between the technologies and crop varieties discussed in the previous two chapters, and the kind of technologies that are the outputs of GCP research. Rather than providing solutions for agricultural problems, the GCP provides a pool of upstream genomics knowledge, capacity and research tools to allow different types of downstream research partners to develop their own solutions. The most convincing and extreme example of this approach is the Genotyping Support Service, which very explicitly does not provide an agricultural product, but a service that can be plugged into different types of research programmes, depending on local needs.

As such, the GSS in particular demonstrates how upstream genomics knowledge can be made accessible upon demand, and cannot replace essential parts of bottom up participatory work, but can be an important complementary element in developing relevant new crop varieties for resource poor farmers. This also creates a situation in which bottom-up and top-down innovation systems are not conflicting or contradictory, but potentially complementary. The availability of a pool of genomics information and genotyping facilities would then create an interface between upstream genomics information and local participatory innovation systems that allows its use in a locally defined and tailored way.

It is important to stress that this positioning of the Generation Challenge Programme, or the Genotyping Support Service does not in itself solve all dilemmas regarding the local adaptation of agro-technological innovation. As mentioned before, the downstream research partners that the GCP collaborates with are of extreme importance for the further development and application of genetic tools made available by GCP research. The observation has been made that GCP is capable of collaborating with a wide variety of downstream research partners, both in terms of the institutional linkages it creates, as well as in terms of its technological outputs. This at least illustrates the practical complementarity of upstream genetic research with various kinds of local organisations involved in agricultural development. However, as mentioned before in this chapter, in order to actually contribute to poverty alleviation, GCP is required to go further than making sure that partnerships are made with downstream partners, and that they are able to take up the outputs of upstream genomics research and translate them into new crop varieties. In addition, it requires the programme to very consciously evaluate who the farmers are that are targeted by the downstream research partners, and are therefore most likely to benefit from the GCP products.

For example, CGIAR institutes and programmes (like the Challenge Programmes) commonly work with the institutes of the national agricultural research systems (NARS) in different developing countries. However, these institutes may have different organisational mandates or objectives than those of the CGIAR or GCP. For example, while the GCP aims to contribute to food security among resource poor farmers in developing countries, national research institutes may have a stronger focus on increasing national food productivity or the revenues from agricultural exports. In practice, these objectives can lead to research programmes for relatively large scale farmers, well connected to modern agricultural inputs, in the most favourable regions of a country. In other words, they do not necessarily target the most resource poor farmers of their country, but the farmers with highest return on investment for their research. While this may be fair enough from a national perspective, if this happens it would imply a disconnect from the objectives to focus agricultural development on poverty alleviation. Similarly, even if coherence in institutional mandates and objectives is not the problem, not all downstream research partners may have the affinity or capacity to set up participatory development projects. So, in summary, the institutional objectives and capacities of downstream research partners remain crucial for the final adaptation and application of new crop varieties or genetic technologies.

In spite of such comments and reservations, the Generation Challenge Programme does provide a valuable example of how diversity and multiplicity in innovation trajectories can take shape, based upon the combination of explorative horizontal and focused vertical research projects, the collaboration with a diverse set of downstream research partners, and by using the Genotyping Support Service as technical interface to connect upstream genomics research data to a wide range of bottom-up research projects. This approach does not reduce the importance of a careful selection of downstream research partners. In addition, it may make demonstrating the exact impacts of GCP research more difficult; an aspect that will be further discussed in the final chapter of this thesis. However, in terms of linking genomics research to development objectives, the networked nature of the Generation Challenge Programme, and the service-like character of the Genotyping Support Service may just be the institutional arrangements that allow upstream genomics science to meaningfully impact upon a multitude of downstream research partners contributing to developing world agriculture.

Chapter 7
Discussing the diversity in approaches to agro-technological innovation

> *"Creative adaptation [...] is not simply a matter of adjusting the form or recoding the practice to soften the impact of modernity; rather, it points to the manifold ways in which a people question the present. It is the site where a people 'make' themselves modern, as opposed to being 'made' modern by alien and impersonal forces, and where they give themselves an identity and a destiny."*
>
> (Gaonkar 2001b, p. 18)

Introduction

The conceptual discussions in Chapter 3, and especially the case studies in Chapter 4 to 6 have provided quite a lot of food for thought regarding possible approaches to agro-technological innovation. This final chapter will look back upon the case studies and aims to provide a synthesis of the research findings and their conceptual and practical implications. In order to do so, it will provide a comparative analysis of the three case studies presented in Chapters 4 to 6, and will discuss the conceptual and practical implications of their findings for contemporary innovation policy in the context of international agricultural development. First, however, the main topic of discussion will be reintroduced, as well as the key concepts that have been used for the case study analysis.

Recapitulation of the study's focus and key concepts

The research presented in this thesis was motivated by an interest in the development of appropriate genetic technologies for agricultural development for resource poor farmers. The introductory chapter elaborated why genetics and genomics may be expected to have the technical potential to contribute to the development of new and useful crop varieties for farmers around the world. However, at the same time the question was raised to what extent technology development not only addresses a specific problem in agricultural production, but also reflects a very specific approach to agricultural development, taking important assumptions on board regarding the production system in which new crop varieties will have to perform and the roles and responsibilities of farmers in that production system. In other words, agricultural development is emphatically studied as not only a technical issue, but as a profoundly social and political issue.

This general research focus led to two major research questions that have been guiding in the case study analysis in this study. These questions – as more extensively introduced and discussed in Chapter 2 – are:

- *How do contemporary projects of pro-poor agricultural biotechnology development operationalize their pro-poor focus, and what criteria are – implicitly and explicitly – taken on board in that consideration?*
- *How are farmers conceptualized as end-users in the operationalization of appropriateness, and what does that mean for their involvement in the innovation process and for their position in the future production process?*

These research questions were further operationalized through a set of study questions (see Chapter 2), which explicitly focused on the institutional context in which agro-technology development was taking place, and the underlying (often implicit) perspectives on the future model of agricultural production.

As elaborated in Chapter 3, this questioning of the changing social relations in agricultural innovation and production has a background in critical studies of biotechnology and agricultural modernisation, that go back to the late 1980s. Studies of Goodman *et al.* (1987) were introduced that highlighted two main organizing principles of agricultural industrialisation: appropriationism and substitutionism. These elements refer to the gradual externalization of aspects of agricultural production, and to the interchangeability of agricultural products, or even their replacement by chemical substitutes. Especially the mechanism of appropriationism highlights how the notion of what it is to be a farmer is fundamentally changing in an industrializing agricultural system. While farmers traditionally have been responsible for experimental breeding, variety management, soil fertility and pest management, many of these elements have been externalized and converted into industrial inputs for a farmer. This implies important changes in the degree to which farmers have access to their means of production, as well to the wider infrastructure that farming becomes a part of, and is dependent upon.

This notion of externalization was further refined by the end of Chapter 3 and a distinction has been made between externalization of elements in the *production system*, and externalization of innovation capacity in an *innovation system*. While the first refers to the decreasing control of the means of production by farmers, the second refers to the degree in which farmers are involved in innovation processes. These different approaches to agricultural innovation were summarized in the dichotomy between farmers that are treated as 'recipients of technology' or farmers that are treated as 'co-innovators' and are hence more prominently involved in the innovation system.

Against the background of these externalization processes, an interest was expressed in the possibility to use genetic technologies for agricultural development, without necessarily externalizing agricultural innovation to specialized breeding institutes or companies, or without transforming farmers' means of production into industrial inputs for farming. This focus was legitimated by the observation that the externalization of agricultural innovation capacity and the means of production was correlated to an increased homogenization in farming styles. While this may be perfectly legitimate in some contexts, a tension was perceived between a standardized package of agricultural advice and technologies, and farming systems in difficult environments which are characterised by a high degree of variability and localized adaptation. As mentioned in Chapter 3, and illustrated by the uneven distribution of the productivity gains of the Green Revolution (Chapter 1), farmers in such areas are likely to require a more open-ended approach to agro-technological innovation in which they play an important role themselves in developing and evaluating new crop varieties and other agricultural technologies. Involving farmers as co-innovators – rather than as recipients of technology – is then expected to create an innovation process that is better capable of dealing with the micro-scale diversity in farming systems and conditions, increasing the capacity of farmers or other local stakeholders to deal with their problems in cultivation themselves. This argument led to the tentative formulation in Chapter 3 of an alternative model of agricultural development, that would challenge the two dimensions of externalization here described. Concretely, this implied challenging the externalization of seed breeding by a consideration of how farmers or grassroots initiatives themselves can be empowered in their varietal management and seed breeding experiments. In addition, external development efforts can be challenged – and complemented – by initiatives of endogenous development (Van der Ploeg and Long 1994) or 'tailor-made biotechnology development' (Ruivenkamp 2003b, 2005), that focus on increasing local innovation capacity, rather than on providing externally developed generic technical solutions to problems in agricultural production.

From the study's rationale to the concrete research focus

The extent to which such underlying trends of externalization in agricultural development are actively and explicitly questioned and lead to reconsiderations in terms of technology development, has been investigated by studying three illustrative cases of contemporary agro-technological development for resource poor farmers (Chapters 4 to 6). A number of concepts have been leading in the study of these cases. First, a notion of 'reflexive development' was introduced, which has been defined as a process of reflecting upon and responding to the effects of development efforts and the comments and criticisms it invokes (Nederveen Pieterse 1998). This concept provided a starting point for a case study analysis that was not only focused on the different approaches between different projects, but more emphatically on the dynamics of institutional learning and change that are taking place and that allow innovative projects from different traditions and backgrounds to reflect upon their own role and position in an agricultural innovation system. So, rather than measuring degrees of reflexivity, the concept provided the case studies with a starting point and a general initial orientation.

Secondly, the concept of 'appropriate technology development' has been introduced, which similarly provided an entry point for the case study analysis. As has been stressed before, this notion of 'appropriate development' is emphatically not defined in a strict or conclusive way. Instead, the ambiguity of the concept is acknowledged and used to tease out the various ways in which it is operationalized in different projects. This studying of how different projects interpret what 'appropriate technology' is, was expected to provide insight in both the explicit and implicit objectives of a project, about its priorities, assumptions and intended end users, and finally about the perspective of how genetic technologies can create an added value in the agricultural production of resource poor farmers in developing countries. Importantly, the notion of appropriateness is studied in the context of 'pro-poor innovation', which also is a concept that is likely to be constructed in different ways in different projects. Therefore, the discussion of how appropriateness is interpreted and operationalized in various ways implicitly also relates to the question of how the qualification 'pro-poor' is put in practice. As mentioned at the end of Chapter 3, the concept of a pro-poor orientation of a development project is guiding for the notion of 'appropriateness', since it defines *for whom* the project or technology should be appropriate. Therefore, both concepts can best be understood in combination, rather than in isolation. Hence, to put it more precisely, the case study analysis focuses on the interpretation of the notion of appropriate technology development *for resource poor farmers.*

Finally, one of the conceptual expectations for this research was that the social and political nature of agricultural development is not only reflected in approaches, strategies and policies for development, but also in the material design of new technologies. This idea is captured in the notion of the 'politics of technological design' as discussed in Chapter 3. More concretely, this notion implies that specific assumptions regarding the future of farming may very literally become embedded in and reflected by the material design of technological artefacts. These assumptions may be related to the specific conditions in which a new crop variety will have to perform, and to local pests and diseases, but also to what kind of farming system is being supported, and what the roles of different social actors are in an agricultural innovation and production system. As such, this exploration of the relationship between technical design and social meaning opens the door to a more profound questioning of technological artefacts themselves.

A brief overview of the three case studies

These concepts here reintroduced have been central in the analysis of the three case studies presented in Chapters 4 to 6. As discussed in Chapter 2, these cases represented different institutional contexts in which pro-poor agro-technological development takes place, both in terms of their public or private nature of funding, as well as in terms of their focus on upstream research or downstream product development. They may not necessarily be representative for the entire range of contemporary projects in which genetic technologies are being used or developed for agricultural development. However, they do provide a rich diversity in contexts, rationales, mandates and approaches that fuel the discussion in this chapter over the various interpretations and operationalizations of appropriate technology development for resource poor farmers. This section provides a very brief reintroduction of the three cases studied and their main characteristics.

First, the case of the 'Collaboration on Insect Management for Brassicas in Asia and Africa' (CIMBAA) has been presented in Chapter 4. This project is a public private partnership involving a range of Indian and international public sector research institutes, and Nunhems Seed (subsidiary of Bayer CropScience) as a private sector company. The consortium is involved in the development of transgenic cabbage varieties that – through the expression of two Bt genes – are intended to provide effective resistance against the diamondback moth. This insect is currently causing major losses in Indian cabbage cultivation, and leads to the application of high doses of insecticides. The development of a transgenic crop variety requires the project to be involved in both the upstream development and testing of new gene constructs, as well as the downstream development and testing of new crop varieties. Vegetable farmers in India are generally small scale farmers, and the consortium is trying to take that context into account in the process of technology development. Moreover, the project has an additional objective in demonstrating that transgenic technology can be used in an effective, safe and socially responsible way, in spite of all the controversy that this technology has raised globally in the past 15 years. As discussed in Chapter 4, these considerations in terms of reaching small scale farmers and in demonstrating the merits of transgenic crops are reflected in a range of considerations and adaptations of the technology that is developed, as well as in the commercialization strategy that is adopted by the project.

Secondly, Chapter 5 focused on the work of the Centro Internacional de la Papa (CIP, International Potato Centre), a research and breeding institute of the Consultative Group on International Agricultural Research (CGIAR). In contrast to the CIMBAA case, the work of CIP is entirely publicly funded. The contrast this provides is not so relevant for the source of funding, as it is for the ambition of the technology developer to gain a commercially interesting position in the future production system that is supported. While this ambition is clearly present in the CIMBAA case study, this does not play a role for CIP. A similarity between both cases is their strong element of downstream product development, and hence their direct interaction with farmers and other stakeholders.

Concretely, the CIP case study focused on the efforts of the institute to develop new potato varieties without jeopardizing the existing diversity that exists in the potato's Andean centre of origin. In practice, this materializes in efforts to widen the output of breeding programmes by releasing a wider variety of clones to farmers that participate in variety selection, and in efforts to add market value to the cultivation of mixtures of potato landraces in the High Andes. In addition, a project on the development of virus resistance kits has been described and analysed in terms of its capacity to empower farmers in their own on-farm production of seed potatoes. Finally, the representation of genetic fingerprinting data in Kipu-like diagrams was discussed as an example of how the institute aims to produce outputs that are not only useful in terms of their technical functioning, but also acceptable in terms of their social meaning and cultural connotations.

Thirdly and finally, Chapter 6 described and discussed the work of the Generation Challenge Programme (GCP), one of the four 'Challenge Programmes' of the CGIAR. The Generation Challenge Programme is committed to the use of comparative genomics, marker assisted breeding, and genotyping technologies to improve plant breeding for resource poor farmers. In particular, the programme focuses on genetic mechanisms underpinning drought tolerance in a wide range of crops. In similarity with CIP, the work of GCP is entirely publicly funded. In contrast, however, the Generation Challenge Programme has a much stronger focus on upstream genomics research in order to elucidate genetic mechanisms for drought tolerance. This focus on upstream research is not exclusive, considering the programme's 'vertical projects' which take a much more focused approach to solving specific problems in agricultural production and reach down to concrete variety development in a specific environment. However, in spite of these vertical projects, the position of the work of GCP vis-à-vis the wider agricultural innovation system is markedly different compared to the work of CIP and CIMBAA.

The study of the GCP has focused on the challenge to link upstream genomics research with the objective to contribute to poverty alleviation among resource poor farmers in drought prone areas. This had led to a discussion on the way GCP has set its research priorities and to what extent these priorities are coherent with the programme's objectives. This in turn led to a reflection upon the strengths and weaknesses of a largely science-led research programme, and the potential complementarities between different types of innovative activity in an innovation system. Especially the Genotyping Support Service was discussed as an example of how technology can be treated as a service, stimulating the interaction between upstream research activities and their implementation in a variety of different downstream development activities.

These three case studies presented in this thesis represent three rather different approaches to technological development for resource poor farmers. The following section will more profoundly engage in a comparative analysis of these cases, based upon the conceptual framework developed in Chapter 3 and briefly reintroduced in the previous section. Later, this chapter will reflect upon the conceptual implications of the case study material, and the implications for contemporary development and innovation policy.

Comparative analysis of the cases – Multiple dimensions of appropriateness

Three case studies have been presented in this thesis, and all three were shown to explicitly reconsider what it means to develop genetic plant breeding technologies for resource poor farmers in developing countries. The comparative analysis in this section will focus on the various interpretations of the notion of appropriate technology development that have provided the rationale for this reflexive process of learning and adaptation. At this point it is important to stress – as has been stressed before – that this thesis does not evaluate the projects studied in terms of their technical success or efficiency. It does not discuss whether – for example – transgenic Bt resistance is a good way to deal with insect infestation, whether participatory breeding approaches can reduce the genetic erosion in the Peruvian Andes, or whether upstream genomics research can successfully contribute to agricultural production in drought prone areas. Rather than getting engaged in such instrumental evaluations, and in spite of their validity, this thesis takes the projects as examples of different ways of engaging with technological development. Rather than their exact outcomes, and whether they are successful in technical terms, their approach, considerations and implications for the social structure of innovation processes are the main objects of analysis for this thesis.

The main research questions of this study question how the studied projects operationalize their 'pro-poor' focus and interpret what it means to provide 'appropriate' technologies for agricultural development. The upcoming sections elaborate what – based upon the case studies – the contours are of a contemporary and practical understanding of appropriate technology development in the context of pro-poor innovation. Concretely, the following dimensions of the notion of appropriateness can be distinguished and will be discussed below:

- Appropriateness in terms of technical functioning.
- Appropriateness with respect to the externalization of the means of production.
- Appropriateness with respect to increasing local innovation capacity.
- Appropriateness in terms of the public perception and cultural connotations of a technology and/or technology developer.

These dimensions are further elaborated in the upcoming sections, and lead to the formulation of a contemporary understanding of the concept appropriateness.

Appropriateness in terms of technical functioning

The first dimension in which the appropriateness of technological innovation is commonly understood is in terms of the *technical functioning* of gene designs, new crop varieties or genotyping technologies in a concrete setting. Technological adaptation must respond to specific local priorities and conditions that may require different technical configurations. This is the most instrumental level of understanding appropriateness of technology in a local context of application, and although it appears straightforward, it is crucial and may be very challenging indeed.

The case studies provided a number of concrete examples of this level of making their technological innovations appropriate. First of all, the concrete technologies or crop varieties that are being developed are responding to specific objectives. Examples include the use of two Bt genes in order to develop effective and durable resistance against the Diamondback Moth infestation (Chapter 4), the development of effective resistance to local virus strains (Chapter 5), and the identification of genetic mechanisms that are responsible for drought tolerance in crops (Chapter 6). However, in addition to the material design of new technologies or crop varieties, the social organisation of innovation processes can play a crucial role in the technical functioning of such innovations. The clearest example is the use of participatory selection trials by CIP in order to identify potatoes with the best characteristics for local production systems (Chapter 4). Although such participatory methodologies can be argued to have different functions, at least one of them is to increase – in functional terms – the fit between the new crop varieties developed, and the preferences of farmers.

These examples illustrate the work that goes into making technological innovations functionally appropriate in a given setting. In a very similar vein, the instrumental appropriateness of innovations is measured and considered in socio-economic terms. In other words: is the technology affordable and remunerative for resource poor, small scale farmers? Especially the CIMBAA case study (Chapter 4) illustrated how socio-economic considerations are reflected in both the technical design of the gene construct, as well as the commercialization strategy. The use of a dual Bt-gene construct theoretically eliminated the need for sowing non-resistant refugia. This in turn means that the technology becomes equally attractive for small scale farmers, who no longer have to sacrifice a significant part of their holdings to a non-resistant crop. Similarly, the efforts to keep the seed price low by not raising a technology fee and by stimulating a diversification within the seed market, are an expression of the objective to reach resource poor farmers with this technology. Although these considerations relate to the social context in which a new crop variety will be marketed, they primarily reflect technical and functional objectives in terms of reaching the intended beneficiaries and in terms of providing an effective solution to problems in agricultural production.

At this point it should be stressed that, although in this section an analytical distinction has been made between the material adaptation of breeding technologies or crop varieties and the social adaptation of innovation systems or commercialization strategies, the case studies have emphasised their interrelationship in practice. In fact, it is the combination of such adaptations that leads to a functional appropriateness in the three case studies. Similarly, sometimes it is the embedding of specific new traits (like virus resistance) within the specific context of a wider technological package (like diagnostic kits) that allows technological innovation to work for resource poor farmers. This observation stresses the importance of considering new technologies, their function and their meaning within the wider context in which they are developed and applied.

The functional adaptation of technological development to a specific context and specific end users is the first – and perhaps most visible – dimension in which the notion of appropriateness is observed to play a role in practice. However, if this instrumental interpretation of appropriateness is used to compare the three case studies, no major differences in their approaches come to the fore. Clearly, the three cases respond differently to the respective contexts in which they are working, to their intended beneficiaries and to the technical challenges they face. However, all three share a general commitment to making their technologies functionally appropriate for resource poor farmers, which is reflected by their reconsiderations of their technical outputs, commercialization strategies and participatory methodologies. This interpretation of appropriateness of technological innovation is therefore important to evaluate the degree to which a project succeeds in providing a functionally suitable and effective solution to a given agricultural problem, but it is not very helpful in distinguishing between the various approaches to agro-technological development that underlie the three case studies – as will become evident from the upcoming sections.

Appropriateness with respect to the externalization of the means of production

As discussed before in Chapter 1 and 3 of this thesis, the understanding of the appropriateness of processes of technological development needs to go beyond a mere evaluation of the technical functioning of new technologies and the infrastructures in which they are introduced. Therefore, questions were raised regarding the externalization of seed breeding and multiplication, and hence regarding the structuring role of specific new crop varieties in production systems. The illustrative example of hybrid maize was mentioned as a classic case in which the material design of a new crop variety fundamentally changed the agricultural production system.

It is at this level that some important differences do become visible between the projects studied. The CIMBAA project was argued to have a strong commitment to reaching small scale vegetable farmers in India; an aim that was reflected by its strategic use of a dual Bt-gene construct. However, while the discussion on the release of open-pollinated varieties did represent a far going willingness to rethink its role as technology developer, the project does implicitly aim at re-establishing the legitimacy and crucial importance of a seed company as provider of high quality cabbage seed, and as provider of solutions for pest management. In fact, the specific trend that the project supports is one in which a precise control over the seed becomes important because of biosafety regulations, as is reflected by the planned release of cytoplasmic male sterile hybrids. As discussed in Chapter 4, this implies that seed companies with the most advanced genetic facilities will have a competitive advantage over seed companies without such facilities. Hence, the technical solution provided by the CIMBAA project may be very useful and effective, but only within the rather industrialized model of vegetable production that already exists in India. This model of agricultural production itself remains unchallenged.

The CIP case study provided some contrasting elements, and most clearly with the example of the development of virus resistance kits. This project relies on the combined use of virus resistant potato varieties, cheap diagnostic tests and improved selection processes to allow the on-farm production of seed potatoes with low levels of virus infection. This may not only allow for an improved management of viruses during cultivation, but importantly has the potential to empower farmers in their own on-farm production of seed potatoes, and reduces their dependence upon externally available commercial virus free seed potatoes. The innovative potential of virus resistance traits here is not restricted to the technical functioning of a potato variety, but is – at least partly – capable of challenging the externalization of seed management in potato production.

These examples illustrate that plant breeding technologies do not only have a technical function within a given production system, but can indeed have an important social or political meaning in challenging or reconfirming a specific production system. In fact, while both the CIMBAA and CIP case studies aim to make their technological outputs functionally appropriate for resource poor farmers, important differences emerge in their approaches to developing seed systems. Rather than arguing that one approach is necessarily better than the other, it is important to acknowledge the completely different settings in which these approaches are chosen. In these settings the preferences of farmers are of relevance, as well as the economics of the existing production systems, and the biology of the different crops that are cultivated. For example, if the on-farm seed production of vegetables (like cabbages) is economically not remunerative or practically impossible in specific agro-climatic conditions, the use of hybrids and the externalization of seed production may be a logical and appropriate choice. In contrast, the clonal propagation of potatoes makes the on-farm production of seed potatoes an interesting alternative, inviting a different approach to seed system development. Finally, the different commercial interests of the publicly funded International Potato Centre and the Nunhems Seed company that is part of the CIMBAA consortium should be considered as an important factor in the kind of production system that is supported. While CIP has no interest in becoming a part of the future potato production system, this may be the single most important motivation for Nunhems to participate in the CIMBAA project. This suggests that the consolidation of an industrial production system is typical for an institutional setting in which a private sector seed company is playing a central role, whereas the more emancipatory approach followed by CIP is typical for an institutional setting in which such commercial considerations do not play a significant role.

In conclusion, next to a technical instrumental interpretation of appropriateness, the concept can be evaluated with respect to the degree in which farmers are gaining or loosing control over (or access to) the means of production, as part of a process of agricultural development. This insight does not directly lead to an argument for which trajectory of agricultural development is most appropriate in general. Instead, the argument is that appropriateness in these terms strongly depends on the context in which a technology is developed, and who its intended end users are. However, this more profound understanding of the notion of appropriateness does allow us to bring certain differences between approaches to technological development to the surface, and to actively question what kind of approach would be most appropriate for agricultural development in a given context.

Appropriateness with respect to increasing local innovation capacity

So far, the discussion on the structuring role of plant breeding technologies has exclusively focused on the CIMBAA and CIP case studies, leaving the Generation Challenge Programme unmentioned. The only reason for this is that such an analysis requires a specific technological artefact within a context of application in order to make sense. As argued in the introduction to Chapter 6, the more upstream work of the Generation Challenge Programme prevents such an analysis, and shifts focus to the linkage between upstream genomics research and downstream development objectives. However, the GCP *is* highly relevant in a comparison of the case studies on a third level of analysis, related to changing roles in the agricultural innovation system.

Whereas the previous section focused on the externalization of seed management in the *production system*, the discussion here relates to externalization in the *innovation system*, determined by the degree to which downstream research partners, or end users can play a more – or less – significant role in determining the exact technical functioning and social meaning of the technology. In this context a dichotomy between treating farmers as 'recipients of technology' or as 'co-innovators' was introduced as a heuristic for the different social roles in an innovation process (Chapter 3). Both terms refer to different approaches to engaging with agricultural and technical development, and with the conceptualization of farmers (or other local stakeholders) in that process. This distinction between different approaches to the process of innovation itself provides a third dimension to questioning the appropriateness of contemporary technology development projects. Although the degree of externalization of innovation processes may be a rather unusual criterion to distinguish between technology development projects, it does provide a new perspective on the case studies presented in this thesis and brings important differences between them sharply in focus. Roughly speaking, whereas the CIMBAA consortium attempted to provide a very concrete solution to a given problem based upon a specific conviction and understanding of the problem and the best solution, the work of CIP and especially of GCP allows for more initiative of farmers and other potential research partners.

The CIMBAA consortium clearly aims to address all concerns that are voiced regarding transgenic technology, and to make the product suitable for small-scale resource poor Indian vegetable farmers. However, the definition of what constitutes an appropriate solution and how it can be developed is primarily formulated by the technology developer. End users and other stakeholders have relatively little influence on the interpretation of the problem, potential solutions, and the concrete technological design that CIMBAA works on. Moreover, the choice for a transgenic insect resistance makes openness in the innovation process rather difficult, for two reasons. The first is that the development of transgenic crops is still highly controversial, making a technology developer vulnerable to attacks from critics of the technology. From the perspective of the technology developer, and notably the private seed company in the consortium, this is perceived as a risk in terms of reputation damage. For that reason, a more defensive approach is being followed in which a product is almost entirely developed and tested, before a wider discussion on its merits is started. Secondly, the use of transgenic technology implies that a project has to meet a lot of regulatory requirements in order to guarantee biosafety within the project. The resulting strict formalization of the innovation process also precludes a more open innovation process in which farmers are involved at an early stage to contribute to experimentation and selection of new cabbage varieties. It is for these reasons that the innovation process is rather closed in nature, and that the project prescribes and supports a specific and arguably narrow understanding of, and solution to the pest infestation problem in cabbages.

In contrast, CIP provides concrete new potato varieties, but also methodologies and practices to allow farmers themselves to keep selecting and cultivating a wide range of potato varieties. The use of participatory variety selection trials in which a wider range of potato varieties is being tested and released illustrates that farmers are consciously involved in the innovation process because of their specific capacities to select the potatoes that are most appropriate within their production systems. This makes the role of CIP in providing a set of potato varieties to be tested still very important, but it means that the innovation process remains somewhat open-ended, and explicitly aimed at local adaptation and selection. This approach is most clearly reflected in the organisation of CIP's breeding programmes, but has a material basis as well, in the sense that the output of the innovation process is no longer a single, widely performing potato variety, but a *collection* of varieties that can be used differently in different regions.

The Generation Challenge Programme, and especially the Genotyping Support Service goes a step further in creating an open-ended innovation process with a crucial role for local stakeholders. The GCP in general produces an upstream pool of genomics information, methodologies and services, which in some vertical projects are fully developed into concrete innovations. However, specifically the Genotyping Support Service inspires a radical reinterpretation of what it means to provide 'appropriate technology' for local agricultural development. If the service is indeed proving to be capable of making a genetic perspective on breeding problems available for local breeding programmes, it would strongly empower their local innovation capacity. The assumption underlying such a service is no longer that a scientific institute can map priorities, translate those into a research programme, and provide a technical solution to a given problem in production. Instead, it would allow local research programmes to approach agricultural problems in a highly contextualized way, and merely provide the technical infrastructure to make use of a genetic perspective in breeding. The focus of technological programmes for agricultural development would then shift from providing a technical product, to providing a technical service. The accessibility of such a service for resource poor farmers communities and the accessibility of the research information feeding into it then become the crucial institutional and regulatory challenges in order to make such an approach work in practice.

These differences in the approaches of the three case studies demonstrate that appropriateness of technological innovation can be interpreted in yet another dimension. This dimension – related to the degree of externalization of the innovation process – is especially relevant against a discussion held in Chapter 3 on the potential for a kind of agricultural development that would allow (and require) farmers to play a significant role in experimentation and evaluation. The comparison of the case studies illustrates that in practice already different approaches are being tried in which farmers play a more significant role in the innovation process, but also that – in other projects – the notion of stakeholder involvement is much less self-evident.

Again, it is important to stress that this analysis does not necessarily provide a clear cut argument for what approach to innovation would be most appropriate in general. Moreover, in order to do the case studies justice, it is important to stress the potential complementarities between different kinds of innovation processes and the different dynamics that have been found within the case studies with respect to the open-endedness of their innovation processes. Especially the Generation Challenge Programme capitalizes upon the complementarity between upstream 'horizontal' research projects and downstream 'vertical' research projects that exhibit different degrees of stakeholder involvement, and interaction with different types of stakeholders. Finally, the degree to which a process of technological innovation may be open-ended and may allow the involvement of various stakeholders depends to a large extent on the nature and objectives of the project. The development of a new crop variety can hardly be as open-ended as the provision of a genotyping methodology, which is not aimed at the development of a concrete technological artefact. However, in spite of such considerations, the idea that a process of innovation is not only about the production of a concrete product, but may in fact be about the creation of innovation capacity for local farming communities, does open the door to a new way of questioning the appropriateness of technological development for resource poor farmers.

Appropriateness in terms of public perception and cultural connotations

The previous sections have discussed the notion of appropriateness of technological innovation in terms of technical functioning, in terms of the externalization in production systems, and in terms of the externalization of innovation processes. This section adds a fourth and final dimension to the notion of appropriateness that has emerged from the CIP case study in particular, and to some extent from the CIMBAA case study as well. This fourth dimension is related to the public perception or cultural connotations of a new technology and the degree to which a technology developer can influence that perception or can respond to it.

This dimension of considering the appropriateness of technology development was most prominently discussed in Chapter 5, with reference to the initiative of CIP to publish genotyping data of native potato varieties in a Kipu-like diagram, which bears strong cultural connotations to pre-Columbian society. While the functional understanding of the genotyping data does not change, and most probably neither does the accessibility of the data by non-scientists, the representation of the data in a Kipu-like diagram was interpreted as an attempt to visually hybridize the world of genetic data with the cultural heritage of indigenous communities in Peru. More important than the question to what extent this attempt is effective, or anything more than a creative attempt to market the scientific outputs of CIP research, its significance lies in the implicit acknowledgement that research outputs not only have a technical function, but also a social and cultural interpretation that may influence their appropriateness for specific communities.

A similar observation has influenced CIP's decision to refrain from working with transgenic material in Peru. Although other arguments against the development of transgenic potatoes existed as well, the understanding that this technology might seriously jeopardize public support and legitimacy of the research institute has been a important factor in the decision to not invest in such research. Interestingly, although similar arguments exist in the context of transgenics development in India, the CIMBAA consortium has consciously decided to work with transgenic cabbages. However, also in this project, the public perception of transgenics plays a crucial role in the decisions the consortium makes. While the consortium willingly confronts public concerns with respect to the development of transgenics, it very consciously tries to address those concerns and allows them to influence the project even if they are not significantly backed up by technical arguments. For example, the safety testing of the transgenic cabbages is explicitly outsourced to public sector research partners in order to avoid the suspicion that the private sector seed company may influence the outcomes of such research because of their commercial interests. This motivation for this decision was the building of trust and credibility for the safety testing, rather than different technical capacities of the institutes that could perform these tests. Similarly, the consideration to release open-pollinated varieties of the transgenic cabbage was primarily a reaction to general public complaints about the power of seed companies to demand new seed purchase every year. The fact that seed production by farmers in India is not remunerative (according to the seed company) and highly uncommon argued against investing in open-pollinated varieties, but the consortium was – in principle – willing to go along with these public concerns, until the point that biosafety regulations effectively prohibited the release of such open-pollinated transgenic varieties.

This dimension of the appropriateness of technology development does not bring new significant differences between the case studies to the surface, even though some projects have to deal with these public perceptions a lot more than others. In general, all three projects studied were perceptive and responsive to this dimension of their work, and acknowledged the importance of avoiding public controversy related to their outputs. A longitudinal study might reveal that this has changed over the 15 years, and that this aspect of making technological development appropriate did not play such a crucial role before the widespread controversy over biotechnology in the late 1990s. However, this study can only observe that the three projects shared a general sensitivity to the public perception of biotechnology and plant breeding, which was reflected in their work.

In conclusion – A contemporary understanding of appropriateness

The previous sections provided a parallel discussion of the various dimensions of appropriateness, as well as a comparative analysis of the three case studies. With respect to the latter, a picture emerged of how the three cases operationalized their objective to contribute to pro-poor agricultural development, in which the most significant differences are related to (1) the kind of seed system that was developed and the degree of farmers' access to their means of production, and (2) the degree to which the innovation process was open-ended and capable to involve farmers as co-innovators in the process. In other words, depending on the context, the three projects followed different approaches when it comes to helping farmers produce their own seed, or in providing good quality seed by a specialized seed producer, and when it comes to providing a 'technological solution' to a given problem, or providing the methodologies and infrastructure to allow farmers to increase their own innovation capacity.

Importantly, the case studies show that both types of externalization processes are not necessarily the same, and that the challenging of one does not necessarily lead to a challenging of the other as well. For example, empowering farmers in their on-farm potato seed production by the provision of virus resistance kits does not necessarily mean that farmers are in fact increasing their innovative capacity to deal with different types of virus infections in the future. On the other hand, increasing the innovative capacity of local breeding programmes does not automatically mean that farmers will also choose to produce their own seed, rather than obtaining it commercially. This thesis does not provide an argument to consider one of these mechanisms more important than the other. Instead, it suggests that both autonomy in production and development, and increasing local innovation capacity are crucial dimensions for rethinking agricultural development and challenging a homogeneous approach to agricultural industrialisation and modernisation.

Next to providing a comparison of the different approaches to agro-technological development in the three cases, these differences allowed for a deconstruction of the notion of appropriateness itself, and its interpretations in these three projects. Making agro-technological development appropriate for resource poor farmers turns out to be a lot more complicated than technically adapting crop varieties to environmental conditions, or than making sure that seed is affordable. Instead, important questions have come to the surface, regarding the social roles of technology developers and farmers in production and innovation systems. This has several implications for the formulation of a contemporary understanding of appropriateness.

First, as becomes clear from the analysis of the case studies, the degree of appropriateness is not only a characteristic of a specific technological artefact or methodology, but equally relates to the characteristics of the innovation process itself. This shifts attention from the outputs of innovation processes, to the dynamics and social relations in an innovation system itself. Secondly, if technological innovation is not only about the provision of technical artefacts, but may be about the provision of a technical service or methodology, this suggests a highly dynamic and interactive relationship with farmers (or other end users of the technology) and technology developers, in which local capacity building continuously leads to new technological requirements. This suggests that the appropriateness of this relationship is a highly dynamic characteristic, rather than a static description of the relationship between a new technology and its intended context of application. Thirdly, the multi-dimensionality of the notion of appropriateness implies that the merits and risks of technological innovation are so dependent upon the perspective on agricultural development that the setup and shaping of a technology development project can never be left to a technology developer alone (or to any other stakeholder for that matter). In fact, this suggests that appropriateness is not achieved by a perfect adaptation and tailoring of any technological artefact to any given situation, but instead requires a continuous reflexivity and interaction between technology developers and end-users; between breeders and farmers.

In summary, appropriateness is probably best understood and treated as a dynamic and continuously changing process characteristic, rather than as a static description of any technological artefact vis-à-vis a fixed context of application. Upcoming sections will further reflect upon this understanding of the notion of appropriateness, its implications for different social roles in agricultural innovation systems, and its relationship with the notion of reflexive (bio)technology development. As a start, the following section will further reflect upon the notion of the 'openness' of innovation, which was identified as an important dimension to consider the appropriateness of technological innovation in.

Additional reflections on the openness of innovation

The elaboration of the various dimensions of appropriateness in the previous sections led to the conclusion that the two major dimensions in which the case studies exhibit differences in their approaches are in terms of the seed systems they support (and the autonomy of farmers in those seed systems), and the degree to which innovation is externalized, or to which farmers can be involved as co-innovators. The first observation builds on the analysis of industrial production systems as has been introduced and elaborated in Chapter 3. Interestingly, the case studies presented in this thesis illustrate that – implicitly – different perspectives on this process of agricultural industrialisation and externalization of the means of production underlie contemporary projects of technology development for resource poor farmers. While those different approaches may be entirely legitimate in different contexts, it is important to bring these differences to the surface in order to be able to discuss and evaluate them.

On a more conceptual level, it is worthwhile to elaborate a bit more on the second observation, regarding the externalization of innovation processes. Even more so since the notion of involving farmers as co-innovators was mentioned as a crucial element in an agricultural innovation system that would be especially suited for farmers in marginal and high risk farming conditions (as discussed in Chapter 3). The comparison of the case studies demonstrated how projects can be distinguished with respect to the degree they treat innovation as an external process (with farmers as recipients of technology) or as an open-ended process (with farmers as co-innovators). In this section and further, the terms 'open innovation' and 'closed innovation' will be used to indicate these different innovation dynamics. An open innovation process is then defined as one that treats and involves farmers as co-innovators and has an open-ended outcome which allows for considerable diversity in local contextualization. A contrasting closed innovation process is defined as a process that produces – in its extreme – a black box which can be used and applied by various end-users, but has little room for local adaptation. Both concepts are considered the extreme ends on a continuum of different innovation styles. They are intended to indicate general trends in the approach to innovation, rather than to represent mutually exclusive ideal types of how innovation is taking place in practice.

This understanding of different innovation dynamics with a varying degree of 'openness' of innovation resonates with other conceptualizations of innovation processes and technological change. For a clear understanding of the concept, it may be helpful to briefly position it vis-à-vis these other ideas. First of all, the term relates to the field of innovation studies in which the term 'open innovation' signifies a strategy in which collaboration, networking and sharing of knowledge is more important for a company, than operating in isolation (Chesbrough 2003). An important aspect in this strategy is the integration of externally initiated and developed competences and creativity within the company itself, in order to use it in the company's innovation processes. This explicitly includes the use of competences and creativity by end-users (such as farmers); a mechanism that has been elaborated by Von Hippel (2005). At the same time, the notion of 'openness' engages with the constructivist notion of 'closure', which leads to a stabilization in technological design. In a constructivist account of technological development, different social groups may have different interpretations of what a technological artefact is for, and hence what the most appropriate technical design is. This flexibility and diversity in technological trajectories ends in a process of stabilization in which a dominant technical design emerges, which then becomes an 'exemplar' for further development in its field (Pinch and Bijker 1984; Van den Belt and Rip 1990). Importantly, such closure or stabilization creates a 'black-box' of the technology, the design of which is no longer called into question but is taken for granted (Feenberg 1999). By implication, in order to meaningfully incorporate the creativity and competences of end-users in an innovation process, their involvement ideally needs to take place before 'closure' of the technical design has taken place.[103]

The observation that different approaches to technological innovation can be described in terms of their 'openness' both incorporates the notion of active end user involvement (a key element of 'open innovation' as described by Von Hippel (2005)), as well as the notion that 'closure' of technical design is postponed or challenged (in the constructivist understanding of technological development). The question now is what the implications of a more open approach to innovation would be for the role of technology developers in an innovation process.

Providing a technical solution, or a technical service

Different approaches to technological development are expected to require or lead to different roles of technology developers and end users. This section explores that mechanism by analysing a shared ambition of the three projects to be perceived and acknowledged as a service provider. Moreover, this discussion will further illustrate the usefulness of the notion of the 'openness of innovation' as entry point to discuss differences between approaches to agro-technological development.

The notion of being a service provider implies that the legitimacy of the technology developer does not depend on the production of a single technology, but on the capacity to bring together technologies and actors to address a wide range of agricultural problems. While this approach is very clear in the case of the Generation Challenge Programme, and to some extent in the plant breeding activities of CIP, it may be less immediately clear how the CIMBAA consortium

[103] At the same time, another process may be identified in which – after closure – technologies are adapted by users and given new meanings or functions. This means that closure may never be final, but always susceptible to modifications in the hands of users. However, returning to the character of the innovation process, this would not exactly be an expression of 'openness in innovation'. In summary, even an externalized and 'closed' innovation process may allow the modification and redesign of technologies in the hands of users, but the interest here goes out to an innovation process that itself is capable of involving users or other stakeholders before closure takes place.

is also presenting itself as primarily a service provider, or rather, a *solution* provider. However, this ambition becomes quite clear from the following interview excerpt:

> *"We have committed ourselves to stopping this project if in a couple of years it turns out that we are not providing the best possible solution. In that case, the best solution [to the problem of insect infestation] should prevail. And as Nunhems [seed company] we then hope to at least have built a reputation as solution provider. Even if we won't have a product, we will have contributed to making the solution visible. And if the solution is one that does not use transgenic technology, but relies on integrated or biological pest control, that is fine as well, as long as we can keep on selling our varieties."*
>
> (anonymous Nunhems representative, interview May 2006)

What this quote demonstrates is the relative detachment of the technical production that CIMBAA is developing, but a strong association with the role of *solution provider*, which legitimates the role of the consortium, and by extension the role of the seed company in the future agricultural production system.

This positioning of the technology developer as a solution provider may be symptomatic for a wider trend in economic development. For example, in their book 'Empire', Hardt and Negri (2000) argue that in the contemporary service-based economy, 'immaterial labour' is of increasing importance, which they define as *"labor that produces an immaterial good, such as a service, a cultural product, knowledge, or communication"* (Hardt and Negri 2000, p. 290). A common element in the different projects then is that in addition to the material labour of producing a specific product, the immaterial labour of producing knowledge, a service, and a beneficial image as solution provider is of a crucial importance. At the same time, on a more practical level, the positioning as a service/solution provider can also be interpreted as a response to criticisms of an overly prescriptive and homogeneous development approach which in the past leaned heavily on the transfer of technology from developed to less-developed countries. Presenting oneself as a solution or service provider can be interpreted as a response to that criticism and as a commitment to making technology development problem-oriented, instead of technology-driven, and flexible to local needs and circumstances.

However, while the service-like nature of the innovation process on the one hand indicates a common element in the different projects, there is clearly a difference between being a 'service provider' or a 'solution provider'. In fact, the shift to a more service-like approach to innovation and production can imply both an externalization of innovation, as well as a contextualization of innovation. This difference is reflected in the approaches adopted by CIMBAA (solution provider) and – for example – the Generation Challenge Programme (service provider), as discussed in the section before. While the ambition to provide a solution in the case of CIMBAA turned out to lead to a rather externalized innovation process, the desire to provide a service implies that the shape of agricultural development is determined by the user of that service, and hence by local breeding projects. In other words, a remarkable analogy between these different approaches lies in their attempts to legitimize the role of the technology developer, beyond the technical product alone, and by their positioning as a resource base which can be helpful in addressing a wide range of problems. However, the way the innovation process is shaped in practice reveals important differences between these approaches, as expressed in terms of the 'openness' of the innovation process.

Flexibility in the relationship between technological design and social meaning

There is a conceptual question that has not been addressed yet in this chapter. The research- and study questions for this study not only inquired into the operationalization of the notions of 'pro-poor development' and 'appropriate technology', but also questioned how different approaches to agricultural development would be reflected by – or embedded in – the concrete technological artefacts that are developed and applied in these projects. This interest in the material basis of technological development was further elaborated in Chapter 3. The point was raised that technologies may embed social norms and configurations in their material design, which gives them a coercive force in prescribing or structuring social organisation. Based upon a conceptual review, a preliminary assumption was made that technological design matters in terms of the social relations of production systems, but that the 'politics of technological design' only come to life within a specific context. The question now is whether the different approaches described in the previous section are also reflected in different material designs of technology. In other words: to what extent did the different approaches to technological development also require different technologies?

The case studies presented in previous chapters show a diverse picture in this respect. While many differences in approaches definitely did have their material basis, in some cases this is much harder to argue. The examples in which a strong relation was visible between the material design of the technology and their social meaning within a production system included the dual Bt gene design (Chapter 4), and the virus resistance kits (Chapter 5). In summary, the dual Bt design explicitly included small-scale farmers as potential end users of the technology, and the virus resistant potato varieties supported farmers as producers of seed potatoes. At the same time, it is difficult to argue that – for example – virus resistance itself has such a profound innate social meaning. As already mentioned in the discussion to Chapter 5, the use of virus resistant potatoes in order to empower farmers in their on farm seed potato production is an example in which this meaning only comes to life in the context of a combined use with virus diagnostic kits and improved selection procedures, and in a socio-economic context in which the commercial prices of seed potatoes are a significant hurdle for potato farmers to replenish their seed stock. The use of virus resistant potatoes in such a project may require no specific technical modification of the potato varieties, but depends on an innovative use and embedding of an already existing technology. Similarly, the material basis underlying CIP's breeding strategies is relevant, but only in a specific context. More than breeding entirely different potatoes, it is an innovative breeding strategy that most clearly reflects a strategy of agricultural development that aims to build on crop genetic diversity rather than to replace it. The material basis of this strategy is not captured in the precise characteristics of a single potato variety, but in the fact that the output is no longer a single variety suited for industrial production, but a collection of potato varieties, which leaves their evaluation and further selection up to farmers. Thirdly, the Generation Challenge Programme and especially the Genotyping Support Service completely fall beyond the scope of such questions regarding the materiality of technologies, at least considering their non-material outputs such as methodologies, knowledge and a service. The focus on drought tolerance as a specific trait was argued (by GCP) to introduce a specific bias for farming systems of resource poor farmers. However, the point was raised that in practice this could not guarantee that resource poor farmers are in fact the main beneficiaries of this new technology, considering that the trait may theoretically also open up arid areas for industrial production systems. Finally, consider the example of molecular fingerprinting data that are represented as a Kipu diagram as described in Chapter 5. In that case, rather than a material redesign of the technology, it is merely a specific symbolic embedding of a genotyping technology that is harnessed to influence the public perception of the technology in the context of indigenous communities in Peru.

It is these examples that stress the importance of a contextualized understanding of technologies and their social meaning in production and innovation systems. This is largely in line with the conceptual starting points as sketched in the section on the politics of technological design in Chapter 3, stressing that technologies can be profoundly political, but that their political meaning depends upon the social and historical context they are part of, and the rituals and discourses they are surrounded by. The practical consequence for agro-technological development is that technologies developed in and for an entirely different context of application cannot be simply adopted uncritically, but also that technological principles that have proven to be successful in an industrial setting may in fact be appropriated in a useful way to become instrumental in quite different models of agricultural production. In summary, this thesis provides additional evidence against an essentialist understanding of the relationship between technical design and social meaning within a production system. The conclusion that can be drawn is that next to focusing on the potential for reconstruction of technologies on a variety of levels, additional attention should go out to acts of 'creative appropriation', which essentially refers to the act of reinventing existing technologies through innovative applications (Feenberg 1999).

Implications of a genetic perspective for stakeholder involvement

Still, in spite of this relativist conclusion regarding the relationship between technical design and social meaning, some concerns may be raised in terms of the use of genetic technologies and the implications for the involvement of farmers or other stakeholders. The question that arises is whether in general an increasing importance of a genetic perspective in plant breeding may lead to a devaluation of the expertise and capabilities of farmers to engage in breeding and selection, and therefore to a further externalization of the breeding process? A more general reflection upon the role of a genetic perspective in breeding may be helpful to address that question.

Clearly, the use of advanced genetic technologies in breeding programmes requires scientific expertise and infrastructure, increasing the importance of geneticists and potentially decreasing the influence and importance of 'genetically unskilled' actors such as farmers or breeders without a specific education in modern genetic technologies and research methods. At the same time, participatory breeding methodologies are a known and proven way of involving farmers in plant breeding, in spite of a more scientist-led upstream research phase (Almekinders and Hardon 2006). But the notion of participatory breeding is not entirely self-evident, nor is its implication in practice. As discussed in Chapter 5, there are different modes of participatory breeding, varying from completely participatory, in which both scientists and farmers are involved in all stages of the breeding process, to mere participatory varietal selection in which farmers are only involved in the downstream evaluation and selection of new varieties. In between both models, an 'efficient participatory breeding' model was proposed which involved farmers in the selection of interesting parental material for crosses, and in the selection of new varieties, but which left the pre-breeding and cultivar development phases up to the scientific breeder (Morris and Bellon 2004). The precise involvement of farmers in these different stages depends on the crop at hand, and especially the extent to which farmers can practically get involved in making crosses. However, in general, the selection of interesting parental material and the selection of best performing varieties are phases in which farmers are expected to bring specific expertise and tacit knowledge to the table, which can improve the breeding programme (*ibid.*).

In that respect, it is important to note that the use of genetic tools in breeding runs a risk of restricting and devaluating these interactions between scientific breeders and farmers. First, marker assisted selection allows for the selection on traits that are invisible to the eye, like durable horizontal resistance, or quality traits related to the biochemical composition of a crop (e.g. the sugar content of potatoes). This means that it can make breeding programmes more efficient and may allow for the breeding of traits that were previously very difficult to select for. However, at the same time it means that – for these traits at least – there is little point in involving farmers in the selection, since they cannot possibly select for such traits. Secondly, on the level of selecting interesting parents for crosses, a genetic perspective is of increasing importance. New genomics technologies allow for the screening of genetic variation in a specific population, which may not be visible to the eye.[104] This means that rather than phenotypical variation in which farmers are arguably experts, genotypic variation is becoming of increasing importance for future breeding programmes, in determining what are interesting parents for crosses.

The implications of these trends on the level of selecting interesting parents for breeding, and selecting the best varieties, are not entirely clear. While on the one hand the privileged and specific knowledge of farmers of traditional varieties may be increasingly replaced by scientific knowledge of the genetic diversity of this germplasm, both perspectives need not be so competing. While marker assisted selection on some traits may be crucial in trait development and pre-breeding, farmer based variety selection may focus on an entirely different set of traits that is harder to quantify, and are more prone to variations because of environmental conditions. In fact, some research suggests that genetic diversity (as mapped by molecular markers) is in fact a rather bad indicator of the phenotypic quality as experienced by farmers (Polycarpe Kayode *et al.* 2006). This argues for a complementarity of marker assisted selection, and farmer mediated selection practices, rather than their competition.

The conclusion must be that a genetic perspective on breeding does not necessarily lead to an externalization of varietal selection. Rather than making farmers' knowledge and experience in parent- and variety selection redundant, both this perspective and the scientific genetic perspective can be highly complementary in breeding programmes. Hence, the challenge lies in finding the institutional and technological configurations in which an increasingly refined genetic perspective can become integrated in bottom-up and pro-poor plant breeding initiatives in order to exploit the complementarity between marker assisted selection on the one hand and farmer based selection on the other. In that respect, fingerprinting technologies for mapping existing biodiversity, and to facilitate marker assisted breeding in a wide range of projects, are examples of what may be considered as 'bridging technologies' that can bring both perspectives on plant breeding together and can make the 'Gene Revolution' interesting for resource poor farmers.

Implications for innovation policy and questions for future research

The discussion of the case studies and the concepts used to analyze them has led to a reflection upon different approaches to technological innovation. This reflection has shifted the attention

---

[104] An example of this is the recently developed technique of 'tilling' (Targeting Induced Local Lesions IN Genomes), a 'reverse genetics' technology designed to detect mutations in plants treated with a mutagen. An extension of 'tilling' is 'ecotilling', in which germplasm collections are screened for genetic variations in a specific gene of interest, without the prior use of a mutagen. This allows the rapid discovery of naturally occurring genetic variation, which may provide interesting leads in the discovery of new gene variations (alleles) of interest (Comai *et al.* 2004; Jackson 2004).

from appropriate technology development in terms of instrumental technical adaptations, to questions regarding the different roles and relationships in a production- and innovation system. While this is to a large extent a conceptual discussion on what it means to develop appropriate technology, it does have some practical implications for agricultural innovation policy. This section discusses those policy implications and in addition introduces some new questions for future research

Conceptual contributions to the debate on alternative agricultural development

The formal aim of this thesis – as formulated in Chapter 2 – is to deepen debates on agricultural development by reflecting upon the relationship between contemporary technology development projects, and a wider context of agricultural modernisation and industrialisation. In fact, this is an ambition to contribute to the contemporary debate on agricultural development on the conceptual level, allowing for a more refined formulation of what is at stake in this process. With respect to this aim, the conceptual discussions in the previous sections have provided a much richer understanding of what it might mean to develop technologies or crop varieties that are 'appropriate' for resource poor farmers. Moreover, the discussion has provided a number of criteria (externalization in production- and innovation systems) that allow for a more articulated and nuanced discussion on the differences between contemporary approaches to agro-technological development and plant breeding.

In addition, a more programmatic aim has been implicit in this thesis, especially in the discussion in Chapter 3 on a potential 'positive alternative' to mainstream agricultural industrialisation and modernisation. An interest was expressed in notions such as 'food sovereignty' (Rosset 2003; Rosset 2006), 'tailormade biotechnologies' (Ruivenkamp 2003b, 2005), 'endogenous development' (Van der Ploeg and Long 1994), and farmer seed systems as 'unsupervised learning networks' (Richards *et al.* 2009). Although different in their language and precise objectives, these concepts share a general concern regarding the appropriateness of mainstream agricultural development for resource poor farmers, especially in marginalized and high-risk farming environments. They question not only the technical adaptation of new technologies to a given context, but also the process and nature of innovation itself. Or, in other words, they not only addresses how agriculture can be improved as quickly and efficiently as possible, but emphatically question the kind of agricultural modernity that is being created and the technologies that play a role in that modernity. A common ground between these critical ideas on agricultural development was found in their concern over externalization processes, and their interest in maintaining (or increasing) farmer autonomy, and local (farmer) innovation capacity. Notably in the case of Richards *et al.* (2009), this led to a direct interest in how novel genetic technologies could contribute to the capacity of farming communities to be directly involved in breeding and selection work.

This thesis has been able to put some flesh on the bones of the idea that technological innovation may be oriented at increasing local innovation capacity and autonomy in production, rather than on providing generic technical solutions. Specifically, it has provided some concrete examples of how genetic fingerprinting technologies or new crop varieties may in practice play an important role in empowering farmers in their on-farm seed production or in their experimentation in local breeding programmes. Moreover, it has demonstrated and discussed that a genetic perspective in plant breeding does not necessarily lead to an externalization of agricultural innovation, but that genetic selection technologies and farmer-based selection can be highly complementary. This concretely contributes to the ongoing work of rethinking the potential role of genetic technologies in agricultural innovation for resource poor farmers.

Maximising and measuring the impact of research

Next to the conceptual and programmatic contributions of this thesis, it also raises some new questions for which it cannot provide a complete answer. For example, much of the ongoing debate on international agricultural research deals with the question of how to increase and measure the impacts of research. The need of having a clear and measurable impact is not only important for companies with commercial interests, but also for public sector agricultural research institution – like those from the CGIAR – that have to legitimize the funding they receive from an international donor committee. That this hasn't exactly been easy for the CGIAR is illustrated by repeated discussions over its impact and its institutional structure (Hall *et al.* 2004b; Alston *et al.* 2006; CGIAR Secretariat 2008). The resulting reform processes in the CGIAR have focused on linking with a wider range of partners, becoming more demand driven, and on institutional efficiency. Recently, this has resulted in a lot of attention for institutional learning and change within the CGIAR (Watts *et al.* 2003; Hall *et al.* 2005; CGIAR 2008).

This thesis has not been directly aimed at discussing how impacts in agricultural research can be maximized. However, it does have some implications for this discussion on maximizing and measuring the impacts of agricultural research. While improving the efficiency of research investments is considered a highly valid objective, the diversity in innovation approaches and dynamics described in this thesis argue for a better alignment of approaches and objectives, rather than for a stronger focus on increasing research impacts per se. In fact, a rather profound contradiction may arise between an innovation policy that is aimed at having (and demonstrating) a maximal impact – as may be required by any research funding body – and an approach to innovation that explicitly leaves open how end users are taking new technologies, methodologies or services on board. While the first requires clear outputs and preferably a wide adoption of those outputs, the second aims to increase local innovation capacity. The latter does not necessarily depend on the provision of a concrete technical artefact, but may in fact depend on the availability of a new technical service or on assistance in local capacity building. So, this thesis does not argue against increasing the impact of agricultural research, but the notion of organizing innovation in an open-ended way – aimed at providing technological 'services', rather that 'solutions' – does raise some important questions regarding the measurement and quantification of those research impacts.

The question that arises is whether a focus on institutional learning and change – as has been dominant in recent reforms in the CGIAR – can also redefine how impacts are measured and what aspects are treated as indicators for success. From an innovation systems perspective the building of linkages between various research partners and end users is emphasised, and treated as both a very important outcome as well as a precondition for innovation processes (Hall *et al.* 2004c, 2005). This argues for measuring impact in terms of institutional learning and change, and in terms of institutional relationships that have been created. A similar approach to measuring impact would be required to evaluate research investments if technologies are treated as a service. Impacts would then have to be measured in terms of research partners or local initiatives that have made use of a given service, or have participated in the development or evaluation of new crop varieties, rather than in the adoption patterns of a limited set of varieties or in terms of productivity gains. This challenge for measuring the impact of investments in research and development indicates a first area in which future research may provide important contributions. New ways of measuring and quantifying the outcomes of research programmes need to be developed in order to legitimize and to evaluate innovation trajectories that are more open ended in nature. This is in no way intended as a way of avoiding accountability of open innovation approaches, but rather as a reconsideration of what useful indicators are to evaluate the success of contemporary development projects.

Partnering policy

In addition to the measurement of impact, this research provides some thoughts on the partnering policy of research institutes like those of the CGIAR, especially regarding public private partnerships. The merits and risks involved in public private partnerships have been an important subject for debate over the recent years (Rausser *et al.* 2000; Hall 2006; Spielman *et al.* 2007). Important advantages of public private partnerships have been mentioned, like increasing the relevance and practical applicability of research, and exploiting the complementarities in research and commercialization capacities (Byerlee and Fisher 2002; Hall *et al.* 2004b). One of the case studies in this thesis (Chapter 4) has specifically dealt with a public private consortium, and its analysis provides an additional criterion to reflect upon the appropriateness of such public private partnerships in agro-technological development.

The analysis of the work of the CIMBAA consortium may have been critical at points, but the thesis definitely does not argue against such public private partnerships. In line with recent research and literature in the area of innovation studies, such partnerships are conceived as a potentially powerful way to access proprietary technologies, to bundle research capacities and to ensure the application of new research findings in actual production systems. However, as stressed in the comparative analysis of the three cases, the structure of the research consortium and the expectations of the research partners regarding their role in the future production system is correlated to a specific approach to agricultural development that legitimizes those roles. Therefore, if the interest of a seed company in joining a public private partnership lies in the potential of future seed sales, the question should be raised what kind of seed system is thought to be most appropriate for the targeted farming communities in the first place. In the case of vegetable cultivation (like in the CIMBAA case), the commercial provision of high quality hybrid seed may definitely be a useful way forward for farmers; however depending on the crop and prevailing mode of production, this is not always self evident. Making the notion of the externalization of seed production an explicit point of discussion allows for a more balanced and nuanced view on the contexts in which public private partnerships may be a useful model of doing research, and when it seems less appropriate.

It should be noted that this analysis is primarily based upon the CIMBAA case study, and therefore on a situation in which a seed company was the private partner in a public private consortium. Things might be different in public private partnerships in which private partners have different commercial interests, for example in the context of post-harvest technologies. Future research is required in order to investigate examples of such partnerships and to reflect upon the correlation with a specific agricultural production system and the repercussions for the position and autonomy of farmers in that production system.

Reflexive biotechnology development and stakeholder involvement

A final practical implication of this study can be indicated, which is related to the importance of reflexivity in agricultural innovation and to the implications for stakeholder involvement at different levels in the innovation process. The comparative case study analysis in this chapter provided a diverse and multi-dimensional picture of how appropriateness took shape in practice in the three projects studied. This led to the conclusion – as discussed earlier in this chapter – that appropriateness is probably best understood and treated as a dynamic and continuously changing process characteristic, rather than as a static description of any technological artefact vis-à-vis a fixed context of application. Moreover, if innovation is not only about the provision of an 'appropriate' technical artefact, but is in fact about local capacity building and institutional learning, this calls for an interactive and dynamic relationship between end users and developers of technology. This understanding of appropriateness does not only refine the debate of what agricultural innovation is all about, but also argues for the encouragement of reflexive (bio) technology development as a model for agro-technological innovation.

This concept of 'reflexive biotechnology development' is a variation on the notion of 'reflexive technology design' which was introduced at the end of Chapter 3 and defined as *"a specific form of deliberative or participatory technology assessment oriented towards the definition of both the problem and the solution in a reciprocal argumentative exchange between the actors involved in the problem"* (Bos 2008, p. 36). This perspective on organizing technological innovation in a deliberative and interactive way nicely fits a notion of 'reflexive development', which was introduced at the very beginning of this thesis. This notion served as a starting point to study contemporary projects of agro-technological development, focusing on the institutional learning dynamics that take place in such projects. Reflexivity was defined as a process of reflecting upon and responding to the effects of development efforts and the comments and criticisms it invokes (Nederveen Pieterse 1998). Moreover, it highlights how development policy increasingly becomes concerned with the management of development interventions itself (*ibid.*).

These elements – the deliberative nature of innovation, and its reflexivity allowing for a continuous realignment of innovation processes and the context of application – are essential in order to contribute to 'appropriate agro-technology development for resource poor farmers' as described in this thesis. In practice, such a reflexive approach implies a focus on a careful alignment of the needs of farmer communities and innovation approaches in terms of the social relations in production and innovation systems. In addition, it calls for a stronger focus on institutional learning and the building of research relationships in order to maximize and measure impacts of research investments. However, although this reflexive approach to agro-biotechnology development does imply a focus on interaction between users and technology developers, it is not simply a general reaffirmation or extension of the plea for stakeholder involvement in all research activities. In fact, as discussed in the context of the Generation Challenge Programme case study, there may be little room for the practical involvement of farmers on the level of upstream genomics research. Rather than finding ways to involve or represent farmers at this upstream stage in research, it was considered to be more fruitful to acknowledge and exploit the complementarities between upstream science-led research programmes and downstream bottom-up initiatives. This shifted attention from direct stakeholder involvement on all levels, to the institutional configurations and technical outputs that would support such complementarity. This in turn led to an interest in the potential of technology as a service – focused on the increase of local innovation capacity – as a complementary model for innovation, next to the development of concrete technical artefacts. In other words, rather than finding ways to meaningfully involve farmers in upstream genomics research, the goal should be to increase the accessibility of research outputs and to stimulate the flexibility in their downstream application by local development initiatives.

In summary – Some practical recommendations for research managers

From the case study analysis and the conceptual discussions in Chapter 3, some practical implications can be distilled for the debate on agricultural development and agro-technological innovation. These can be summarized in the following brief recommendations for policy makers or research managers in the field of agro-technology development:

- Encourage reflection upon the alignment between research approaches and the model of agricultural development that may be most appropriate for the intended beneficiaries, especially in terms of the production system and innovation system that is created or supported.
- Encourage the experimentation with breeding strategies that focus on the release of a wider set of new varieties and that aim to complement the cultivation of traditional varieties, rather than replace it. Moreover, encourage the experimentation with crop varieties that allow farmers to be involved in their on farm seed production and hence may be instrumental in the strengthening of the informal seed sector. These strategies may not be appropriate for all farmers in all regions, but they provide a potentially valuable alternative to the trends of an externalization of the means of production and the decrease of local innovation capacity, and may as such be highly appropriate for farmers in regions that have so far been left behind in attempts of agricultural development.
- Encourage the exploitation of complementarities between a genetic perspective on plant breeding and the capacities of farmers in variety selection, in order to maximize the relevance of new crop varieties.
- Invest in institutional learning processes and develop methods to measure research impacts along those lines.
- Encourage public private partnerships for the potential complementarity in capacities and the access to proprietary IP, but be careful in the kind of production system that is supported and the interests of the private partners to play a future role as technology provider. The model of agricultural modernisation that is interesting for the private partner may not always be the most appropriate model for resource poor farmers.
- Encourage stakeholder involvement and interactive research processes, but don't blindly extend this effort to upstream genomics research. In that domain, it makes much more sense to reflect on the way in which complementarities with different downstream partners can be exploited, and how new technologies and methodologies can be made available as a service, rather than as a concrete artefact.

In conclusion – Reflexive biotechnology development

Contemporary debates over international agricultural development not only deal with questions over how to achieve agricultural and economic development as fast and efficiently as possible. In addition, the kind of agricultural modernity that is created with the introduction of new farming practices, technologies and institutes is being questioned. The questioning of modernity and the search for alternatives *of*, or alternatives *to* modernity is a wider theme in development debates. Interestingly, one of the crucial elements shared by the critical writings on agricultural development referred to in this thesis, is that they do not aim to create a singular alternative modernity, but essentially treat (agricultural) modernity as something plural, something that is firmly grounded in local culture, economy, and natural environment. This plurality in modernities is elegantly incorporated in a concept used by Dilip Parameshwar

Goankar. He claims that in contrast to a number of clearly visible trends (in terms of global institutional arrangements and financial networks), our contemporary globalizing world is not simply converging to one single global culture, economy, market or science (Gaonkar 2001b). His edited volume 'Alternative Modernities' provides a range of cases that prove his argument by drawing attention to 'creative adaptations' of modernisation processes on a local scale (Gaonkar 2001a). In the words of Gaonkar, already quoted at the beginning of this chapter:

> *"Creative adaptation [...] is not simply a matter of adjusting the form or recoding the practice to soften the impact of modernity; rather, it points to the manifold ways in which a people question the present. It is the site where a people 'make' themselves modern, as opposed to being 'made' modern by alien and impersonal forces, and where they give themselves an identity and a destiny."*
>
> (Gaonkar 2001b, p. 18)

This 'making oneself modern' is a direct parallel to the evaluation of agro-technological development in terms of autonomy in production and self-determination in terms of innovation. In addition, the notion of creative adaptation provides a counter perspective to the notion of 'creative destruction' that was mentioned in Chapter 3 as an essential characteristic of an expanding capitalist system of production. While creative destruction essentially represents the replacement of one production system by another, a process in which farmers are passive objects of change, creative adaptation emphasises the possibility for farmers or other local stakeholders to be the subjects of their own development process and to take existing situations and production systems as starting point for development.

As Gaonkar implies with his notion of 'creative adaptation', it is ultimately local stakeholders – whether they are breeders, farmers or consumers – that are in charge of taking development into their own hands, and in shaping modernity to their own views, needs and desires. Notwithstanding this crucial role of local stakeholders in appropriating (technology) development, this thesis has focused on the role of technology developers in prescribing technological solutions, or in empowering local stakeholders to come up with their own solutions. It has done so by questioning how different technology developers have operationalized the notion of 'appropriateness' in their projects and how that changed the social roles of both technology developers and farmers. The notion of appropriateness served as a vehicle to explore and discuss the relationship between the concrete decisions that any technological project makes in terms of technical design and institutional organisation, and its wider perspective on the 'future of farming' and the role of different stakeholders in that future.

What emerged was a diverse and multi-dimensional picture of how appropriateness took shape in practice in the three projects studied. This contemporary understanding of appropriateness was taken as an argument for reflexive biotechnology development as an approach to agro-technological innovation. This reflexive approach crucially implies an interactive and dynamic relationship between farmers and technology developers, and a more profound reflection upon the social roles of these stakeholders in future production and innovation systems. This outline of reflexive biotechnology development is not intended as a coherent alternative to contemporary mainstream agro-technological development, nor does it necessarily provide a better approach to innovation for all farmers. However, it ís expected to provide some valuable starting points for rethinking and improving agricultural innovation for resource poor farmers in marginal and high risk areas that have been largely left behind by previous efforts to agricultural modernisation. By encouraging a more profound reflection upon and reconsideration of the effects on production and innovation systems, this research may inspire technology developers or policy makers to question their own role in innovation processes. This research will have achieved its objectives if that leads to more technology trajectories which allow for a meaningful integration of farmers, and farmer-led innovation processes into scientific technological development, allowing for the creative adaptation of agricultural technologies in a plural and diverse way, for an infinite range of local agri-cultures.

References

Adi, B. (2006). 'Intellectual property rights in biotechnology and the fate of poor farmers' agriculture.' *The journal of world intellectual property* **9**(1): 91-112.

Akrich, M. (1992). 'The de-scription of technical objects.' In: *Shaping technology/Building society*. W.E. Bijker and J. Law (eds.) Cambridge, Massachusetts: The MIT Press: 205-224.

Almekinders, C.J.M. and J.J. Hardon (eds.) (2006). *Bringing farmers back into breeding. Experiences with participatory plant breeding and challenges for institutionalisation*. Wageningen: Agromisa. Agrospecial 5, 135 p.

Almekinders, C.J.M., G. Thiele and D.L. Danial (2007). 'Can cultivars from participatory plant breeding improve seed provision to small-scale farmers?' *Euphitica* **153**: 363-372.

Alston, J.M., S. Dehmer and P.G. Pardey (2006). 'International initiatives in agricultural R&D: The changing fortunes of the CGIAR.' In: *Agricultural R&D in the developing world: Too little too late?* P.G. Pardey, J.M. Alston and R.R. Piggot (eds.) Washington, DC: IFPRI.

ANDES and IIED (2005). 'Traditional Resources, Rights and Indigenous Peoples in the Andes.' IIED Reclaiming Diversity Series. London: IIED.

Art, H.W. (ed.) (1993). *The Dictionary of ecology and environmental science* New York: Henry Holt & co.

Atkinson, R.C., R.N. Beachy, G. Conway, F.A. Cordova, M.A. Fox, K.A. Holbrook, D.F. Klessig, R.L. McCormick, P.M. McPherson, H.R. Rawlings III, R. Rapson, L.N.Vanderhoef, J.D.Wiley and, C.E.Young (2003). 'Public sector collaboration for agricultural IP management.' *Science* **301**(5630): 174-175. Jul 11.

Beck, U. (1992). *Risk Society: towards a new modernity*. London: Sage.

Becker, H.S. (1993). 'Theory: the necessary evil.' In: *Theory and concepts in qualitative research: Perspectives from the field*. D.J. Flinders and G.E. Mills (eds.) New York: Teachers College Press: 218-229.

Benjamin, W. (2003). *Selected Writings, Volume 4, 1938-1940*. Cambridge & London: Harvard University Press.

Bernstein, H. (1992). 'Agrarian structures and change: India.' In: *Rural livelihoods: crises and responses*. H. Bernstein, B. Crow and H. Johnson (eds.) Oxford: Oxford University Press: 51-64.

Bickman, L., D.J. Rog and T.E. Hedrick (1998). 'Applied research design. A practice approach.' In: *Handbook of applied social research methods*. L. Bickman and D.J. Rog (eds.) London: Sage Publications.

Bijker, W.E. (1995). *Of bicycles, bakelites, and bulbs: toward a theory of sociotechnical change*. Cambridge: MIT Press.

Blumer, H. (1954). 'What's wrong with social theory?' *American sociological review* **18**: 3-10.

Bos, A.P. (2008). 'Instrumentalization Theory and Reflexive Design in Animal Husbandry ' *Social Epistemology* **22**(1): 29-50.

Bowen, G.A. (2006). 'Grounded theory and sensitizing concepts.' *International Journal of Qualitative Methods* **5**(3): 12-23.

Brewster, A.L., A.R. Chapman and S.A. Hansen (2005). 'Facilitating Humanitarian Access to Pharmaceutical and Agricultural Innovation.' *Innovation Strategy Today* **1**(3). 22 Sept 2006.

Broerse, J.E.W. (1998). *Towards a new development strategy. How to include small-scale farmers in the biotechnological innovation process*. Delft: Eburon.

Broerse, J.E.W. and J.F.G. Bunders (2000). 'Requirements for biotechnology development: the necessity for an interactive and participatory innovation process.' *International Journal of Biotechnology* **2**(4): 275-296.

Brooks, S. (2005). 'Biotechnology and the Politics of Truth: From the Green Revolution to an Evergreen Revolution.' *Sociologia Ruralis* **45**(4): 360-379.

Brown, L.R. (1997). 'The agricultural link: how environmental deterioration could disrupt economic progress'. World Watch Paper No.136. Washington, D.C: World Watch Institute.

Brush, S.B. (1992). 'Reconsidering the Green-Revolution – Diversity and Stability in Cradle Areas of Crop Domestication.' *Human Ecology* **20**(2): 145-167. Jun.

Brush, S.B. (2000). 'The issues of in situ conservation of crop genetic resources'. In: *Genes in the field: On-farm conservation of crop diversity*. S.B. Brush (ed.) Rome, Italy: International Plant Genetic Resources Institute.

Brush, S.B. (2003). 'The demise of 'common heritage' and protection for traditional agricultural knowledge'. Conference on Biodiversity, Biotechnology and the Protection of Traditional Knowledge, St. Louis MO. April 4-5, 2003.

Bruskiewich, R., G. Davenport, T. Hazekamp, T. Metz, M. Ruiz, R. Simon, M. Takeya, J. Lee, M. Senger, G. McLaren and T.J.L. van Hintum (2006). 'Generation Challenge Programme (GCP): Standards for Crop Data.' *OMICS: A Journal of Integrative Biology* **10**(2): 215-219.

Bucchini, L. and L.R. Goldman (2002). 'Starlink corn: A risk analysis.' *Environmental Health Perspectives* **110**(1): 5-13. Jan.

Buijs, J., M. Martinet, F. De Mendiburu and M. Ghislain (2005). 'Potential adoption and management of insect-resistant potato in Peru, and implications for genetically engineered potato.' *Environmental Biosafety Research* **4**(3): 179-188.

Bunders, J. (1988). 'Appropriate Biotechnology for Sustainable Agriculture in Developing-Countries.' *Trends in Biotechnology* **6**(8): 173-180. Aug.

Bunders, J.F.G. and J.E.W. Broerse (eds.) (1991). *Appropriate biotechnology in small-scale agriculture: how to reorient research and development*. Wallingford: CAB International.

Buttel, F.H., M. Kenney and J.R. Kloppenburg (1985). 'From Green Revolution to Biorevolution: Some Observations on the Changing Technological Bases of Economic Transformation in the Third World.' *Economic Development and Cultural Change* **34**(1): 31-55.

Byerlee, D. and K. Fisher (2002). 'Accessing modern science: Policy and institutional options for agricultural biotechnology in developing countries.' *World Development* **30**(6): 931-948.

CGIAR (2008). 'ILAC at a glance'. *ILAC Fact Sheet 1; November 2008*.

CGIAR Secretariat (2008). 'A revitalized CGIAR – A new way forward'. AGM 08 Brief,.

Chambers, R. (1994). 'Participatory Rural Appraisal (PRA): Analysis of experience.' *World Development* **22**(9): 1253-1268.

Chambers, R., A. Pacey and L.A. Thrupp (eds.) (1989). *Farmer First: Farmer innovation and agricultural research*. London: Intermediate Technology Publications.

Chataway, J. (2005). 'Introduction: is it possible to create pro-poor agriculture-related biotechnology?' *Journal of international development* **17**: 597-610.

Chaturvedi, S. and L. Chawii (2005). 'Biosafety protocol, international trade and agricultural biotechnology: policy inferences for India. RIS-DP #99'. RIS. http://www.ris.org.in/dp99_pap.pdf (last accessed 17 September 2008).

Chesbrough, H.W. (2003). *Open Innovation: the new imperative for creating and profiting from technology.* Boston, Massachusetts: Harvard Business School Press.

Chiarolla, C. (2006). 'Commodifying agricultural biodiversity and development-related issues.' *The Journal of World Intellectual Property* **9**(1): 25-60.

CIP (2007). 'Transgenic potatoes will not be released by CIP in the Andean countries.' CIP Press Release, 26 July 2007. Online available at: http://www.cipotato.org/pressroom/press_releases_detail.asp?cod=41 (last accessed 17 September 2008).

CIP and CIDA (2006). 'Andean agriculture in the Altiplano Project (ALTAGRO) – First progress report August 2005 – February 2006.' International Potato Center (CIP) & Canadian International Development Agency (CIDA). Online available at: http://inrm.cip.cgiar.org/altagro/docs/report01.pdf (last accessed 17 September 2008).

CIP and FEDECH (2006). *Catálogo de variedades de papa nativa de Huancavelica – Perú.* Lima (Peru): Centro Internacional de la Papa (CIP), Federación Departamental de Comunidades Campesinas (FEDECH).

Comai, L., K. Young, B.J. Till, S.H. Reynolds, E.A. Greene, C.A. Codomo, L.C. Enns, J.E. Johnson, C. Burtner, A.R. Odden and S. Henikoff (2004). 'Efficient discovery of DNA polymorphisms in natural populations by ecotilling.' *Plant Journal* **37**: 778-786.

Conway, G. (1998). *The doubly green revolution: Food for all in the twenty-first century.* Ithaka, NY: Cornell University Press.

Cooke, B. and U. Kothari (eds.) (2001). *Participation: The New Tyranny?* London: Zed.

Cooper, D., R. Vellvé and H. Hobbelink (1992). *Growing Diversity: Genetic Resources and Local Food Security.* London: Intermediate Technology Publications.

Dawea, D., R. Robertson and L. Unnevehr (2002). 'Golden rice: what role could it play in alleviation of vitamin A deficiency?' *Food Policy* **27**(5-6): 541-560.

De Haan, S. and G. Thiele (2005). 'In situ conservation and potato seed systems in the Andes.' In: *Seed systems and crop genetic divesrity on-farm – Proceedings of a Workshop, 16–20 September 2003, Pucallpa, Peru.* D.I. Jarvis, J. L. Chávez-Servia and T. Hodgkin (eds.) Rome: IPGRI.

Deibel, E. (2006). 'Common genomes: open source in biotechnology and the return of common property.' *Tailoring Biotechnologies* **2**(2): 49-84.

Delmer, D.P. (2005). 'Agriculture in the developing world: Connecting innovations in plant research to downstream applications.' *PNAS* **102**(44): 15739-15746.

Devaux, A., C. Velasco, G. López, T. Bernet, M. Ordinola, H. Pico, G. Thiele and D. Horton (2007). 'Collective Action for Innovation and Small Farmer Market Access: The Papa Andina Experience.' CAPRI Working Paper No. 68 – October 2007.

Diao, X., P. Hazell, D. Resnick and J. Thurlow (2007). 'The role of agriculture in development. Implications for Sub-Saharan Africa. Research Report 153.' Washington D.C.: IFPRI.

Dixon, C. (1990). *Rural development in the third world.* London: Routledge.

Dixon, J., A. Gulliver, D. Gibbon and M. Hall (eds.) (2001). *Farming systems and poverty – Improving farmers' livelihoods in a changing world.* Rome and Washington D.C.: FAO and World Bank.

Dorward, A., J. Kydd, J. Morrison and I. Urey (2004). 'A policy agenda for pro-poor agricultural growth.' *World Development* **32**(1): 73-89. Jan.

Douglas, J. (1980). *Successful seed programs: a planning and management guide.* Boulder, Co.: Westview Press.

Dutfield, G. (2003). *Intellectual property rights and the life science industries. A twentieth century history*. Hampshire: Ashgate Publishing Limited.

Edquist, C. (ed.) (1997). *Systems of Innovation. Technologies, institutions and organisations*. Science, technology and the international political economy. London: Pinter.

Edwards, P.N. (2003). 'Infrastructure and Modernity: Force, time, and social organisation in the history of sociotechnical systems.' In: *Modernity and Technology*. T.J. Misa, P. Brey and A. Feenberg (eds.) Cambridge, Massachusetts & London, England: The MIT Press: 185-226.

Escobar, A. (1995). *Encountering development*. Princeton NJ: Princeton University Press.

Evenson, R.E. and D. Gollin (2003). 'Assessing the impact of the Green Revolution, 1960 to 2000.' *Science* **300**(5620): 758-762. May 2.

Falcon, W.P. and C. Fowler (2002). 'Carving up the commons – emergence of a new international regime for germplasm development and transfer.' *Food Policy* **27**(3): 197-222. Jun.

FAO (2001). 'Global farming systems.' Spotlight, June 2001,.

Feenberg, A. (1999). *Questioning technology*. New York: Routledge.

Ferreira, S.A., K.Y. Pitz, R. Manshardt, F. Zee, M. Fitch and D. Gonsalves (2002). 'Virus Coat Protein Transgenic Papaya Provides Practical Control of Papaya ringspot virus in Hawaii.' *Plant Disease* **86**(2): 101-105.

Flick, U. (2006). 'Coding and categorizing.' In: *An introduction to qualitative research*. U. Flick (ed.) London: SAGE.

Ford Runge, C. and B. Senauer (2007). 'How biofuels could starve the poor.' Foreign Affairs,.

Foucault, M. (1977). *Discipline and punish*. New York: Pantheon. Translation: A. Sheridan.

Freebairn, D.K. (1995). 'Did the Green-Revolution Concentrate Incomes – a Quantitative Study of Research Reports.' *World Development* **23**(2): 265-279. Feb.

Freeman, C. (1987). *Technology and economic performance: lessons from Japan*. London: Pinter.

Frewer, L. (2003). 'Societal issues and public attitudes towards genetically modified foods.' *Trends in Food Science & Technology* **14**(5-8): 319-332.

Gaonkar, D.P. (ed.) (2001a). *Alternative Modernities*. Durham & London: Duke University Press.

Gaonkar, D.P. (2001b). 'On alternative modernities.' In: *Alternative modernities*. D.P. Gaonkar (ed.) Durham & London: Duke University Press.

Garst, R. and T. Barry (1990). *Feeding the crisis: U.S. food aid and farm policy in central america*. Lincoln: University of Nebraska Press.

Generation Challenge Programme (2006). 'Generation Challenge Programme – Priorities and Strategy; draft 3 july 2006.' Available online at: http://www.generationcp.org/comm/manual/doc_087.doc (last accessed 17 September 2008).

Generation Challenge Programme (2007). 'Generation Challenge Programme Strategic Framework – 6 Februari 2007.' Available online at: http://www.generationcp.org/comm/gcp_framework_final.pdf (last accessed 17 September 2008).

Gepts, P. (2004). 'Who owns biodiversity, and how should the owners be compensated?' *Plant Physiology* **134**(4): 1295-1307. Apr.

Ghislain, M., A. Lagnaoui and T. Walker (2003). 'Fulfilling the promise of Bt potato in developing countries.' *Journal of New Seeds* **5**: 93-113.

Gisselquist, D., J. Kampen, T. Sykes and G. Alex (1998). 'Initiatives for sustainable seed systems in Africa.' Seed policy and programmes for sub-saharan Africa, Abidjan, Côte d'Ivoire: FAO. 23-27 November 1998.

Glaser, B.G. and A. Strauss (1967). *The discovery of grounded theory: Strategies for qualitative research.* Chicago: Aldine.

Glover, D. (2007). 'Farmer participation in private sector agricultural extension.' *IDS Bulletin* **38**(5): 61-73.

S.A. Goff, D. Ricke, T.-H. Lan, G. Presting, R. Wang, M. Dunn, J. Glazebrook, A. Sessions, P. Oeller, H. Varma, D. Hadley, D. Hutchison, C. Martin, F. Katagiri, B.M. Lange, T. Moughamer, Y. Xia, P. Budworth, J. Zhong, T. Miguel, U. Paszkowski, S. Zhang, M. Colbert, W.-l. Sun, L. Chen, B. Cooper, S. Park, T.C. Wood, L. Mao, P. Quail, R. Wing, R. Dean, Y. Yu, A. Zharkikh, R. Shen, S. Sahasrabudhe, A. Thomas, R. Cannings, A. Gutin, D. Pruss, J. Reid, S. Tavtigian, J. Mitchell, G. Eldredge, T. Scholl, R. M. Miller, S. Bhatnagar, N. Adey, T. Rubano, N. Tusneem, R. Robinson, J. Feldhaus, T. Macalma, A. Oliphant and S. Briggs (2002). 'A draft sequence of the rice genome (Oryza sativa L. ssp. japonica).' *Science* **296**(5565): 92-100. Apr 5.

Goodman, D., B. Sorj and J. Wilkinson (1987). *From farming to biotechnology. A theory of agro-industrial development.* Oxford: Basil Blackwell Ltd.

Graff, G.D., S.E. Cullen, K.J. Bradford, D. Zilberman and A.B. Bennett (2003). 'The public private structure of intellectual property ownership in agricultural biotechnology.' *Nature Biotechnology* **21**(9): 989-995. Sep.

Graff, G.D., B.D. Wright, A.B. Bennett and D. Zilberman (2004). 'Access to intellectual property is an major obstacle to developing transgenic horticultural crops.' *California Agriculture* **58**(2): 120-126.

GRAIN (2005). 'Seed laws: imposing agricultural apartheid – Editorial.' Seedling,.

Grin, J., F. Felix, A.P. Bos and S.F. Spoelstra (2004). 'Practices for reflexive design:Lessons from a Dutch programme on sustainable agriculture.' *International Journal of Foresight and Innovation Policy* **1**(1-2): 126-49.

Grint, K. and S. Woolgar (1995). 'On some failures of nerve in constructivist and feminist analyses of technology.' *Science, technology & human values* **20**(3): 286-310.

Guba, E.G. and Y.S. Lincoln (1989). *Fourth Generation Evaluation.* Sage: Thousand Oaks, London, New Delhi.

Guerinot, M.L. (2000). 'The Green Revolution strikes gold.' *Science* **287**(1): 241-243.

Hall, A. (2005). 'Capacity development for agricultural biotechnology in developing countries: an innovation systems view of what it is and how to develop it.' *Journal of international development* **17**: 611-630.

Hall, A. (2006). 'Public private sector partnerships in an agricultural system of innovation: concepts and challenges.' *UNU-Merit Working papers*(#2006-002). January 2006.

Hall, A., N. Clark, S.V. Rasheed, M.V.K. Sivamohan and B. Yoganand (2000). 'New agendas for agricultural research in developing countries: Policy analysis and institutional implications.' *Knowledge, Technology and Policy* **13**(1): 70-79.

Hall, A., L. Mytelka and B. Oyeyinka (2005). 'Innovation systems: Implications for agricultural policy and practice.' *ILAC Brief 2; July 2005.*

Hall, A., R.V. Sulaiman, B. Yoganand and N. Clark (2003). 'Post-harvest innovation systems in South Asia: research as capacity development and its prospects for pro-poor impact.' *Outlook on Agriculture* **132**(2): 19-26.

Hall, A., R.V. Sulaiman, B. Yoganand, R.S. Raina, N.G. Clark and G.C. Naik (2004a). 'Institutional learning and change: towards a capacity-building agenda for research. A review of recent research on post-harvest innovation systems in South Asia.' In: *Innovations in Innovation: reflections on partnership, institutions and learning.* A. Hall, B. Yoganand, R.V. Sulaiman R.S. Raina, C.S. Prasad, G.C. Naik and N.G. Clark (eds.) Patancheru, New Delhi: CPHP, ICRISAT, NCAP.

Hall, A. and B. Yoganand (2004). 'New institutional arrangements in agricultural research and development in Africa: concepts and case studies'. In: *Innovations in Innovation: reflections on partnership, institutions and learning.* A. Hall, B. Yoganand, R.V. Sulaiman R.S. Raina, C.S. Prasad, G.C. Naik and N.G. Clark (eds.) Patancheru, New Delhi: CPHP, ICRISAT, NCAP.

Hall, A., B. Yoganand, J.H. Crouch and N.G. Clark (2004b). 'The evolving culture of science in the Consultative Group on International Agricultural Research: concepts for building a new architecture of innovation in agri-biotechnology'. In: *Innovations in Innovation: reflections on partnership, institutions and learning.* A. Hall, B. Yoganand, R.V. Sulaiman R.S. Raina, C.S. Prasad, G.C. Naik and N.G. Clark (eds.) Patancheru, New Delhi: CPHP, ICRISAT, NCAP.

Hall, A., B. Yoganand, R.V. Sulaiman R.S. Raina, C.S. Prasad, G.C. Naik and N.G. Clark (eds.) (2004c). *Innovations in Innovation: reflections on partnership, institutions and learning.* Patancheru, New Delhi: CPHP, ICRISAT, NCAP.

Halloy, S.R.P., R. Ortega, K. Yager and A. Seimon (2005). 'Traditional Andean cultivation systems and implications for sustainable land use'. In: *ISHS Acta Horticulturae 670: I International Symposium on Root and Tuber Crops: Food Down Under* CAB Abstracts: 31-55.

Halpin, C. (2005). 'Gene stacking in transgenic plants – the challenge for 21st century plant biotechnology.' *Plant Biotechnol J* **3**(2): 141-55. Mar.

Hardt, M. and A. Negri (2000). *Empire.* Cambridge: Harvard University Press.

Herrera-Estrella, L. (1999). 'Transgenic plants for tropical regions: Some considerations about their development and their transfer to the small farmer.' *Proceedings of the National Academy of Science* **96**(11): 5978-5981.

Herrera, S. (2005). 'Richard Jefferson – Profile.' *Nature Biotechnology* **23**(6): 643.

Hettne, B. (1990). *Development theory and the three worlds.* London: Longman.

Hobbelink, H. (1991). *Biotechnology and the future of world agriculture.* London, New Jersey: Zed Books Ltd.

Hobsbawm, E. (1995). *The age of extremes: The short twentieth century 1914-1991.* London: Little, Brown Book Group.

Hoisington, D., M. Khairallah, T. Reeves, J.-M. Ribaut, B. Skovmand, S. Taba and M. Warburton (1999). 'Plant genetic resources: What can they contribute toward increased crop productivity?' *Proceedings of the National Academy of Sciences of the United States of America* **96**(11): 5937-5943. May 25.

Hope, J.E. (2004). *Open source biotechnology.* Australian National University: PhD Dissertation.

Horton, D. (1987). *Potatoes, production, marketing and programs fro developing countries.* Boulder: Westview Press.

Hughes, S. (2005). 'Navigating genomes: the space in which genes happen.' *Tailoring Biotechnologies* **1**(1): 35-46.

Hyman, G., S. Fujisaka, P. Jones, S. Wood, M. Carmen de Vicente and J. Dixon (2008). 'Strategic approaches to targeting technology generation: Assessing the coincidence of poverty and drought-prone crop production.' *Agricultural Systems* **98**(1): 50-61.

Jackson, M.T. (2004). 'Assessing EcoTILLING as a methodology for targeted genotyping and SNP discovery'. Proposal submitted to the Generation Challenge Program Commissioned Research (SP1), 25 November 2004. Manila, Philippines: IRRI.

Jacobsen, L.E. (1964). 'The Ancient Inca Empire of Peru and the Double Entry Accounting Concept.' *Journal of Accounting Research* **2**(2): 221-228.

Jaillon, O., J. Aury, B. Noel, A. Policriti, C. Clepet, A. Casagrande, N. Choisne, S. Aubourg, N. Vitulo, C. Jubin, A. Vezzi, F. Legeai, P. Hugueney, C. Dasilva, D. Horner, E. Mica, D. Jublot, J. Poulain, C. Bruyère, A. Billault, B. Segurens, M. Gouyvenoux, E. Ugarte, F. Cattonaro, V. Anthouard, V. Vico, C. Del Fabbro, M. Alaux, G. Di Gaspero, V. Dumas, N. Felice, S. Paillard, I. Juman, M. Moroldo, S. Scalabrin, A. Canaguier, I. Le Clainche, G. Malacrida, E. Durand, G. Pesole, V. Laucou, P. Chatelet, D. Merdinoglu, M. Delledonne, M. Pezzotti, A. Lecharny, C. Scarpelli, F. Artiguenave, M.E. Pè, G. Valle, M. Morgante, M. Caboche, A.-F. Adam-Blondon, J. Weissenbach, F. Quétier and P. Wincker (2007). 'The grapevine genome sequence suggests ancestral hexaploidization in major angiosperm phyla.' *Nature* **449**(7161): 417. September 27 2007.

James, C. (2006). 'Global status of commercialized biotech/GM crops: 2006.' ISAAA Brief no. 35. Ithaka, NY: ISAAA.

James, C. (2007). 'Global Status of Commercialized Biotech/GM Crops: 2007 '. ISAAA Brief No. 37. Ithaka, NY: ISAAA.

Jansen, J.P.A. (1997). 'Aphid resistance in composites.' *International application published under the patent cooperation treaty (PCT).*

Jayaraman, K.S. (2001). 'Illegal Bt cotton in India haunts regulators.' *Nature Biotechnology* **19**(12): 1090-1090. Dec.

Jayaraman, K.S. (2004). 'Illegal seeds overtake India's cotton fields.' *Nature Biotechnology* **22**(11): 1333-1334. Nov.

Jefferson, R.A. (2006). 'Science as social enterprise. The CAMBIA BiOS Initiative.' *Innovations*(Fall 2006): 13-44.

Joerges, B. (1999). 'Do politics have artefacts?' *Social Studies of Science* **29**(3): 411-431. Jun.

Johnson, N. and J. Berdegué (2004). 'Property rights, collective action and agribusiness.' In: *Collective action and poverty rights for sustainable development – 2020 Focus No. 11.* IFPRI (ed.) Washington: International Food Policy Research Institute.

Johnson, N.L. and V.W. Ruttan (1994). 'Why are farms so small?' *World Development* **22**(5): 691-706.

Jongerden, J. (2008). 'First the peasant? Some reflections on modernity, technology and reconstruction.' In: *Reconstructing biotechnologies: Critical social analyses.* G. Ruivenkamp, S. Hisano and J. Jongerden (eds.) Wageningen: Wageningen Academic Publishers: 119-142.

Kassawike, K. (2008). *Irrigation-based Livelihood Challenges and Opportunities. A gendered technography of irrigation development intervention in the Lower Moshi irrigation scheme in Tanzania.* Wageningen: PhD thesis, Wageningen University.

Khush, G.S. (1999). 'Green revolution: preparing for the 21st century.' *Genome* **42**: 646-655.

Kirwan, J.R. (2003). *The reconfiguration of producer-consumer relations within alternative strategies in the UK agro-food system: the case of farmers' markets*: PhD thesis, University of Gloucestershire.

Kloppenburg, J.R. (1988). *First the seed. The political economy of plant biotechnology 1492-2000.* Cambridge: Cambridge University Press.

Kloppenburg, J.R. [1988] (2004). *First the seed. The political economy of plant biotechnology. Second edition.* Madison, Wisconsin: The University of Wisconsin Press.

Kloppenburg, J.R. (2004). 'Still the seed: plant biotechnology in the twenty-first century.' In: *First the seed. The political economy of plant biotechnology 1492-2000. Second edition.* Madison, Wisconsin: The University of Wisconsin Press.

Kogan, M. (1998). 'Integrated pest management: Historical perspectives and contemporary developments.' *Annual Review of Entomology* **43**: 243-270.

Koo, B., C. Nottenburg and P.G. Pardey (2004). 'Intellectual property. Plants and intellectual property: an international appraisal.' *Science* **306**(5700): 1295-7. Nov 19.

Kryder, R.D., S. Kowalski and A.F. Krattiger (2000). 'The intellectual and technical property components of pro-Vitamin A rice (GoldenRice^{TM}): A preliminary freedom-to-operate review.' *ISAAA briefs* **20**.

Lammerts van Bueren, E.T. (2006). 'Organic plant breeding: a challenge for practice and science'. Paper presented at Joint Organic Congress, Odense, Denmark. May 30-31 2006 Paper available online at: http://orgprints.org/7857/01/odense_paper_ELammerts_van_Bueren_2006.pdf (last accessed 17 September 2008).

Lammerts van Bueren, E.T. and P.C. Struik (2004). 'The consequences of the concept of naturalness for organic plant breeding and propagation.' *NJAS – Wageningen Journal of Life Sciences* **52**(1): 85-95.

Lammerts van Bueren, E.T., P.C. Struik and E. Jacobsen (2002). 'Ecological concepts in organic farming and their consequences for an organic ideotype.' *Netherands Journal of Agricultural Science* **50**(1): 1-26.

Lander, E.S., L.M. Linton, B. Birren, C. Nusbaum, M.C. Zody, J. Baldwin, K. Devon, K. Dewar, M. Doyle, W. FitzHugh, R. Funke, D. Gage, K. Harris, A. Heaford, J. Howland, L. Kann, J. Lehoczky, R. LeVine, P. McEwan, K. McKernan, J. Meldrim, J.P. Mesirov, C. Miranda, W. Morris, J. Naylor, C. Raymond, M. Rosetti, R. Santos, A. Sheridan, C. Sougnez, N. Stange-Thomann, N. Stojanovic, A. Subramanian, D. Wyman, J. Rogers, J. Sulston, R. Ainscough, S. Beck, D. Bentley, J. Burton, C. Clee, N. Carter, A. Coulson, R. Deadman, P. Deloukas, A. Dunham, I. Dunham, R. Durbin, L. French, D. Grafham, S. Gregory, T. Hubbard, S. Humphray, A. Hunt, M. Jones, C. Lloyd, A. McMurray, L. Matthews, S. Mercer, S. Milne, J.C. Mullikin, A. Mungall, R. Plumb, M. Ross, R. Shownkeen, S. Sims, R.H. Waterston, R.K. Wilson, L.W. Hillier, J.D. McPherson, M.A. Marra, E.R. Mardis, L.A. Fulton, A.T. Chinwalla, K.H. Pepin, W.R. Gish, S.L. Chissoe, M.C. Wendl, K.D. Delehaunty, T.L. Miner, A. Delehaunty, J.B. Kramer, L.L. Cook, R.S. Fulton, D.L. Johnson, P.J. Minx, S.W. Clifton, T. Hawkins, E. Branscomb, P. Predki, P. Richardson, S. Wenning, T. Slezak, N. Doggett, J.F. Cheng, A. Olsen, S. Lucas, C. Elkin, E. Uberbacher, M. Frazier, R.A. Gibbs, D.M. Muzny, S.E. Scherer, J.B. Bouck, E.J. Sodergren, K.C. Worley, C.M. Rives, J.H. Gorrell, M.L. Metzker, S.L. Naylor, R.S. Kucherlapati, D.L. Nelson, G.M. Weinstock, Y. Sakaki, A. Fujiyama, M. Hattori, T. Yada, A. Toyoda, T. Itoh, C. Kawagoe, H. Watanabe, Y. Totoki, T. Taylor, J. Weissenbach, R. Heilig, W. Saurin, F. Artiguenave, P. Brottier, T. Bruls, E. Pelletier, C. Robert, P. Wincker, D.R. Smith, L. Doucette-Stamm, M. Rubenfield, K. Weinstock, H.M. Lee, J. Dubois, A. Rosenthal, M. Platzer, G. Nyakatura, S. Taudien, A. Rump, H. Yang, J. Yu, J. Wang, G. Huang, J. Gu, L. Hood, L. Rowen, A. Madan, S. Qin, R.W. Davis, N.A. Federspiel, A.P. Abola, M.J. Proctor, R.M. Myers, J. Schmutz, M. Dickson, J. Grimwood, D.R. Cox, M.V. Olson, R. Kaul, C. Raymond, N. Shimizu, K. Kawasaki, S. Minoshima, G.A. Evans, M. Athanasiou, R. Schultz, B.A. Roe, F. Chen, H. Pan, J. Ramser, H. Lehrach, R. Reinhardt, W.R. McCombie, M. de la Bastide, N. Dedhia, H. Blöcker, K. Hornischer, G. Nordsiek, R. Agarwala, L. Aravind, J.A. Bailey, A. Bateman, S. Batzoglou, E. Birney, P. Bork, D.G. Brown, C.B. Burge, L. Cerutti, H.C. Chen, D. Church, M. Clamp, R.R. Copley, T. Doerks, S.R. Eddy, E.E. Eichler, T.S. Furey, J. Galagan, J.G. Gilbert, C. Harmon, Y. Hayashizaki, D. Haussler, H. Hermjakob, K. Hokamp, W. Jang, L.S. Johnson, T.A. Jones, S. Kasif, A. Kaspryzk, S. Kennedy, W.J. Kent, P. Kitts, E.V. Koonin, I. Korf, D. Kulp, D. Lancet, T.M. Lowe, A. McLysaght, T. Mikkelsen, J.V. Moran, N. Mulder, V.J. Pollara, C.P. Ponting, G. Schuler, J. Schultz, G. Slater, A.F. Smit, E. Stupka, J. Szustakowski, D. Thierry-Mieg, J. Thierry-Mieg, L. Wagner, J. Wallis, R. Wheeler, A. Williams, Y.I. Wolf, K.H. Wolfe, S.P. Yang, R.F. Yeh, F. Collins, M.S.

Guyer, J. Peterson, A. Felsenfeld, K.A. Wetterstrand, A. Patrinos, M.J. Morgan, J.o.n.g. de, J.J. Catanese, K. Osoegawa, H. Shizuya, S. Choi, Y.J. Chen (2001). 'Initial sequencing and analysis of the human genome.' *Nature* **409**(6822): 860-921. Feb 15.

Latouche, S. (1993). *In the wake of the affluent society: An exploration of post-development.* London: Zed.

Latour, B. (1992). 'Where are the missing masses? The sociology of a few mundane artifacts.' In: *Shaping technology/Building society.* W.E. Bijker and J. Law (eds.) Cambridge, Massachusetts: The MIT Press: 225-258.

Latour, B. (2004). 'Which politics for which artifacts?' Retrieved 17 September 2008, from http://www.bruno-latour.fr/presse/presse_art/GB-06%20DOMUS%2006-04.html.

Le Buanec, B. (2004). 'Protection of plant-related innovations: evolution and current discussion.' *IP Strategy Today*(9-2004): 1-18.

Levidow, L. (2008). 'European quality agriculture as an alternative bio-economy'. In: *Reconstructing Biotechnologies: Critical social analyses.* G. Ruivenkamp, S. Hisano and J. Jongerden (eds.) Wageningen: Wageningen Academic Publishers: 185-206.

Lin, Y. and S. Tanaka (2006). 'Ethanol fermentation from biomass resources: current state and prospects.' *Applied microbiology and biotechnology* **69**(6): 627-642.

Louwaars, N. (1994). 'Connecting formal and local seed systems in development: integrated seed supply systems'. In: *Organisation and management of national seed programmes.* P. Stolp (ed.) Aleppo, Syria: DSE/ICARDA, Proceedings of a follow-up seminar/workshop,12-24 November 1994.

Louwaars, N. (1997). 'Regulatory Aspects of Breeding for Field Resistance in Crops.' *Biotechnology and Development Monitor* **33**: 68.

Louwaars, N. (2005). 'Genetically Modified Organisms (GMOs) – Patents, Licenses and Liability'. Seed Info, 29, July 2005,.

Louwaars, N. (2007). *Seeds of Confusion; The impact of policies on seed systems.* Wageningen University: PhD dissertation.

Louwaars, N. and C.J.M. Almekinders (2002). 'The Importance of the Farmers' Seed Systems in a Functional National Seed Sector.' *Journal of New Seeds* **4**(1/2): 15-33.

Lovett, J.C., G.F. Midgley and P. Barnard (2005). 'Climate change and ecology in Africa.' *African Journal of Ecology* **43**(3): 167–169.

Lundvall, B.A. (ed.) (1992). *National Systems of Innovation. Towards a theory of innovation and interactive learning.* London: Pinter.

MacKenzie, D. and J. Wajcman (1985). 'Introduction: The social shaping of technology'. In: *The social shaping of technology.* (First edition) D. Mackenzie and J. Wajcman (eds.) Buckingham, UK: Open University Press.

MacKenzie, D. and J. Wajcman (eds.) (1999). *The social shaping of technology.* Second edition. Buckingham: Open University Press.

Madhavan, M.R. and K. Sanyal (2006). 'PRS Legislative Brief: Seed Bill 2004': India Together. http://www.indiatogether.org/2006/jun/law-seeds.htm (last accessed 17 September 2008).

Marcuse, H. (1964). *One-dimensional man: studies in the ideology of advanced industrial society.* Boston: Beacon Press.

Marris, C., B. Wynne, P. Simmons and S. Weldon (2001). 'Public perceptions of agricultural biotechnologies in Europe'. Final Report of the PABE research project.

Max-Neef, M.A. (1991). *Human-scale Development.* New York: Apex Press.

McKusick, V.A. and F.H. Ruddle (1987). 'Editorial – A new discipline, a new name, a new journal.' *Genomics* **1**(1): 1-2.

Moffat, A. S. (2002). 'Finding new ways to protect drought-stricken plants.' *Science* **296**(5571): 1226-1229.

Mohan, G. (2001). 'Beyond participation: Strategies for deeper empowerment.' In: *Participation: the new tyranny?* B. Cooke and U. Kothari (eds.) London & New York: Zed Books: 153-167.

Mohan, M. and G.T. Gujar (2003). 'Local variation in susceptibility of the diamondback moth, Plutella xylostella (Linnaeus) to insecticides and role of detoxification enzymes.' *Crop Protection* **22**(3): 495-504. Apr.

Momsen, J.H. (1991). *Women and development in the third world.* London: Routledge.

Morley, S.A. (2001). 'Distribution and growth in Latin America in an era of structural reform.' TMD Discussion paper. Washington D.C.: International Food Policy Research Institute. Online available at: http://www.ifpri.org/divs/tmd/dp/papers/tmdp66.pdf (last accessed 17 September 2008).

Morris, M.L. and M.R. Bellon (2004). 'Participatory plant breeding research: Opportunities and challenges for the international crop improvement system.' *Euphytica* **136**(1): 21-35.

Naylor, R.L., W.P. Falcon, R.M. Goodman, M.M. Jahn, T. Sengooba, H. Tefera and R.J. Nelson (2005). 'Biotechnology in the developing world: a case for increased investments in orphan crops.' *Food Policy* **29**: 15-44.

Nederveen Pieterse, J. (1998). 'My paradigm or yours? Alternative development, post-development, reflexive development.' *Development and Change* **29**: 343-373.

Negi, G.C.S. (1994). 'High Yielding vs. Traditional Crop Varieties: A Socio-Agronomic Study in a Himalayan Village in India ' *Mountain Research and Development* **14**(3): 251-254.

Nelson, R.R. (ed.) (1993). *National Innovation Systems: a comparative study.* New York: Oxford University Press.

Opara, L.U. (2003). 'Traceability in agriculture and food supply chain: a review of basic concepts, technological implications, and future prospects.' *Journal of Food, Agriculture and Environment* **1**(1): 101-106.

Ortiz, O. (2006). 'Evolution of agricultural extension and information dissemination in Peru: An historical perspective focusing on potato-related pest control.' *Agriculture and Human Values* **23**(4): 477-489.

Parayil, G. (2003). 'Mapping technological trajectories of the Green Revolution and the Gene Revolution from modernization to globalization.' *Research Policy* **32**(6): 971-990. Jun.

Perkins, J.H. (1997). *Geopolitics and the Green Revolution: wheat, genes, and the cold war.* New York: Oxford University Press.

Pesticide Outlook (2002). 'Top award for scientist who developed the first transgenic plant.' *Pesticide Outlook* **13**(June 2002): 93.

Pfaffenberger, B. (1992). 'Technological Dramas.' *Science Technology & Human Values* **17**(3): 282-312. Sum.

Pickersgill, B. and B. Heiser (1978). 'Origins and distributions of plants domesticated in the New World tropics.' In: *Advances in Andean archaeology.* D.L. Browman (ed.) The Hague: Mouton: 133-165.

Pinch, T.J. and W.E. Bijker (1984). 'The Social Construction of Facts and Artifacts – or How the Sociology of Science and the Sociology of Technology Might Benefit Each Other.' *Social Studies of Science* **14**(3): 399-441.

Pingali, P.L. and P.W. Heisy (1999). 'Cereal crop productivity in developing countries: past trends and future prospects.' CIMMYT Economics paper 99-03. Mexico D.F.: CIMMYT.

Pinstrup-Andersen, P. and M.J. Cohen (2000). 'Modern Biotechnology for Food and Agriculture: Risks and Opportunities for the Poor.' In: *Agricultural Biotechnology and the Poor.* G.J. Persley and M.M. Lantin (eds.) Washington, D.C.: The World Bank.

Pistorius, R. and J. Van Wijk (1999). *The exploitation of plant genetic information. Political strategies in crop development.* Oxon: Cabi Publishing.

Pollan, M. (2001). *The botany of desire: a plant's-eye view of the world.* New York: Random House Inc.

Polycarpe Kayode, A.P., A.R. Linnemann, M.J.R. Nout, J.D. Hounhouigan, T.J. Stomph and M.J.M. Smulders (2006). 'Diversity and food quality properties of farmers' varieties of sorghum from Bénin.' *Journal of the Science of Food and Agriculture* **86**: 1032-1039.

Pretty, J.N. (1995). *Regenerating Agriculture: Policies and Practice for Sustainability and Self-Reliance.* Washington D.C.: Joseph Henry Press.

Pretty, J.N. (2002). *Agri-culture. Reconnecting people, land and nature.* London: Earthscan.

Puente-Rodríguez, D. (2008). 'The Wiphala Genomics: the deployment of molecular markers in small-scale potato crop systems in the Bolivian Andes.' *The European Journal of Development Research* **20**(3): 377-398.

Quaye, W. (2007). 'Food sovereignty and combating poverty and hunger in Ghana.' *Tailoring Biotechnologies* **3**(2): 69-78.

Rausser, G., L. Simon and H. Ameden (2000). 'Public–private alliances in biotechnology: Can they narrow the knowledge gaps between rich and poor?' *Food Policy* **25**: 499–513.

Reardon, T. and C.B. Barrett (2000). 'Agroindustrialization, globalization, and international development: An overview of issues, patterns and determinants.' *Agricultural Economics* **23**: 195-205.

Reece, J.D. and E. Haribabu (2007). 'Genes to feed the world: The weakest link?' *Food Policy* **32**: 459-479.

Ribaut, J.-M. (2006). *Drought Adaptation in Cereals.* New York: Haworth's Food Products Press.

Ribaut, J.-M., P. Monneveux, J.-C. Glaszman, H. Leung, T. Van Hintum and C. de Vicente (2008). 'International Programs and the Use of Modern Biotechnologies for Crop Improvement '. In: *Genomics of Tropical Crop Plants.* P.H. Moore and R. Ming (eds.) New York: Springer 21-61.

Richards, P. (1985). *Indigenous agricultural revolution: ecology and food production in West Africa.* London: Unwin Hyman.

Richards, P., M. De Bruin-Hoekzema, S.G. Hughes, C. Kudadjie-Freeman, S. Kwame Offei, P.C. Struik and A. Zannou (2009). 'Seed systems for African food security: Linking molecular genetic analysis and cultivator knowledge in West Africa.' *International Journal of Technology Management* **45**(1/2): 196-214.

Röling, N. and A. Wagemakers (eds.) (1998). *Facilitating sustainable agriculture. Participatory learning and adaptive management in times of environmental uncertainty.* Cambridge: Cambridge University Press.

Rosegranta, M.W., N. Leacha and R.V. Gerpacioa (1999). 'Alternative futures for world cereal and meat consumption.' *Proceedings of the Nutrition Society* **58**: 219-234.

Rosset, P. (2003). 'Food Sovereignty: global rallying cry of farmer movements.' *Backgrounder* **9**(4): Fall 2003.

Rosset, P. (2006). 'Agrarian reforms and food sovereignty: Alternative model for the rural world.' 55th Annual Conference of the Center for Latin American Studies: Alternative Visions of Development: The Rural Social Movements in Latin America, Gainesville, Florida, USA. February 23-25, 2006.

Ruivenkamp, G. (1986). 'The impact of biotechnology on international development: competition between sugars and sweeteners.' *Vierteljahresberichte des forschungsinstituts der Friedrich Ebert Stiftung*(103, March. Special issue on 'New technologies and Third World development').

Ruivenkamp, G. (1989). 'De invoering van biotechnologie in de agro-industriële productieketen: de overgang naar een nieuwe arbeidsorganisatie.' *Faculteit politieke en sociaal-culturele wetenschappen.* Amsterdam: Universiteit van Amsterdam. PhD thesis: 354.

Ruivenkamp, G. (2003a). 'Biotechnology: the production of new identities.' In: *Organisational relationships in the networking age. The dynamics of identity formation and bonding.* W. Koot and P. Leisink (eds.) Cheltenham: Edward Elgar.

Ruivenkamp, G. (2003b). 'Tailor-made biotechnologies for endogenous developments and the creation of new networks and knowledge means.' *Biotechnology and Development Monitor* **50**: 14-16. Mar.

Ruivenkamp, G. (2005). 'Tailor-made biotechnologies: between bio-power and sub-politics.' *Tailoring Biotechnologies* **1**(1).

Ruttan, V.W. (2004). 'Controversy about agricultural technology: lessons from the green revolution.' *International Journal of Biotechnology* **6**(1): 43-54.

Sachs, J.D. (2005). *The end of poverty: Economic possibilities for our time.* New York: Penguin Press.

Sachs, W. (1992). *The Development Dictionary: A guide to knowledge as power.* London: Zed.

Sahai, S. (2007). 'Cotton economics in India is confused.' BioSpectrum India, 7 March 2007, pp. 66-67.

Sandur, S. (2004). 'Implications of Diamondback moth control for Indian cabbage farmers.' Report for CIMBAA. http://www.cimbaa.org/downloads/CIMBAA_IndianDBM_ReportFinal-%20Apr05.pdf (last accessed 17 September 2008).

Schaart, J.G. (2004). 'Towards consumer-friendly cisgenic strawberries which are less susceptible to Botrytis cinerea.' Wageningen: Wageningen University, the Netherlands.

Schnepf, E., N. Crickmore, J. Van Rie, D. Lereclus, J. Baum, J. Feitelson, D.R. Zeigler and D.H. Dean (1998). 'Bacillus thuringiensis and Its Pesticidal Crystal Proteins.' *Microbiology and Molecular Biology Reviews* **62**(3): 775-806.

Schot, J. (2003). 'The contested rise of a modernist technology politics.' In: *Modernity and Technology.* T.J. Misa, P. Brey and A. Feenberg (eds.) Cambridge, Massachusetts & London, England: The MIT Press: 257-278.

Scott, J.C. (1998). *Seeing like a state. How certain schemes to improve the human condition have failed.* New Haven & London: Yale University Press.

Seedling (2005). 'Jack Kloppenburg – Interview.' Seedling: pp. 18-20.

Sen, A. (1981). *Poverty and famines: An essay on entitlement and deprivation.* Oxford: Clarendon Press.

Sen, A. (1999). *Development as Freedom.* Oxford: Oxford University Press.

Seshia, S. and I. Scoones (2003). 'Tracing policy connections: the politics of knowledge in the green revolution and biotechnology eras in India.' IDS working paper 188, June 2003.

Shelton, A.M., J.-Z. Zhao and R.T. Roush (2002). 'Economic, ecological, food safety, and social consequences of the deployment of Bt transcenic plants.' *Annual Review of Entomoloy* **47**: 845-881.

Shetty, P.K. (2004). 'Socio-ecological implications of pesticide use in India.' Economic and Political Weekly, December 4, 2004, 5261-5267.

Shiva, V. (1991). *The Violence of the Green Revolution. Third world agriculture, ecology and politics.* London & New Jersey: Zed Books Ltd.

Shumacher, E.F. (1973). *Small Is Beautiful: Economics As If People Mattered* London: Blond and Briggs.

Shumpeter, J.A. [1942] (1975). *Capitalism, Socialism, and Democracy.* New York: Harper.

Singh, R.B. (2000). 'Environmental consequences of agricultural development: a case study from the Green Revolution state of Haryana, India ' *Agriculture, Ecosystems & Environment* **82**(1-3): 97-103.

Smale, M. (1997). 'The Green Revolution and Wheat Genetic Diversity: Some Unfounded Assumptions.' *World Development* **25**(8): 1257-1269.

Smits, R. and S. Kuhlman (2004). 'The rise of systemic instruments in innovation policy.' *Int. J. Foresight and Innovation Policy* **1**(1/2): 4-32.

Smolders, W. (2005). 'Plant genetic resources for food and agriculture: facilitated access or utility patents on plant varieties?' *IP Strategy Today*(13-2005): 1-17.

Snapp, S. (1999). 'Mother and baby trials: a novel trial design being tried out in Malawi.' TARGET – The Newsletter of the Soil Fertility Research Network for Maize-Based Cropping Systems in Malawi and Zimbabwe, January 1999 issue – CIMMYT, Zimbabwe,.

Snapp, S.S., M.J. Blackie and C. Donovan (2003). 'Realigning research and extension to focus on farmers' constraints and opportunities.' *Food Policy* **28**(4): 349-363.

Snapp, S.S. and K.L. Heong (2003). 'Scaling up: participatory research and extension to reach more farmers.' In: *Uniting Science and Participation for Adaptive Natural Resource Management*. B. Pound, S.S. Snapp, C. McDougal and A. Braun (eds.) UK & Canada: Earthscan & IRDC: 67-87.

Sperling, L., J.A. Ashby, M.E. Smith, E. Weltzien and S. McGuire (2001). 'A framework for analyzing participatory plant breedingapproac hes and results.' *Euphytica* **122**(3): 439-450.

Spielman, D.J., F. Hartwich and K. Von Grebmer (2007). 'Public–Private Partnerships in International Agricultural Research.' IFPRI, Research Brief No. 9.

Spitz, P. (1987). 'The Green Revolution re-examined in India.' In: *The Green Revolution Revisited: critique and alternatives*. B. Glaeser (eds.) London: Unwin Hyman: 57-75.

Spooner, D.M., K. McLean, G. Ramsay, R. Waugh and G.J. Bryan (2005). 'A single domestication for potato based on multilocus amplified fragment length polymorphism genotyping.' *PNAS* **102**(41): 14694-14699.

Srinivasan, R., N.S. Talekar and V. Dhawan (2005). 'Transgenic plants with dual Bt gene – an innovative initiative for sustainable management of Brassica insect pests.' *Current Science* **88**(12). 25 June 2005.

Strauss, M. (2000). 'When Malthus meets Mendel.' *Foreign Policy* **119**: 105-112.

Sullivan, S.N. (2005). 'The evolving international regime of liability and redress relating to the use of genetically modified organisms: a prelimary report.' In: *Genetic resource policies and the Generation Challenge Programme*. N. Louwaars (ed.) Mexico, D.F.: Generation Challenge Programme.

Swaminathan, R., and M.S. Swaminathan. (2004) 'From Green Revolution to Gene Revolution.' PBI Bulletin, Issue 2: 3-7. Ottawa: National Research Council, Canada Plant Biotechnology Institute.

Swart, J.A.A., J. Jiang and P. Ho (2007). 'Risk perceptions and GM crops: The case of China.' *Tailoring Biotechnologies* **3**(3): 11-28. Winter 2007-2008.

Talekar, N.S. and A.M. Shelton (1993). 'Biology, Ecology, and Management of the Diamondback Moth.' *Annual Review of Entomology* **38**: 275-301.

Taylor, C. (2001). 'Two theories of modernity.' In: *Alternative modernities*. D. Parameshwar Gaonkar (ed.) Durham & London: Duke University Press.

Thiele, G. (1999). 'Informal potato seed systems in the Andes: Why are they important and what should we do with them?' *World Development* **27**(1): 83-99. Jan.

Thiele, G. (2000). 'Bibliography of participatory research in CIP [Bibliografía de investigación participativa en el CIP].' Social Science Department Working Paper No. 2000-3. International Potato Center (CIP), Lima, Peru.

Thiele, G. and A. Devaux (2002). 'Adding value to local knowledge and biodiversity of Andean potato farmers.' SDC project news, 2-3.

Thiele, G., E. v. d. Fliert and D. Campilan (2001a). 'What happened to participatory research at the International Potato Center?' *Agriculture and Human Values* **18**: 429-446.

Thiele, G., R. Nelson, O. Ortiz and S. Sherwood (2001b). 'Participatory Research and Training: Ten Lessons from the Farmer Field Schools (FFS) in the Andes.' *Currents* **27**: 4-11.

Thirtle, C., X. Irz, L. Lin, V. McKenzie-Hill and S. Wiggins (2001). 'Relationship between changes in agricultural productivity and the incidence of poverty in developing countries.' DFID Report No. 7946 27/02/2001.

Thomas, A. (2000). 'Poverty and the 'end of development''. In: *Poverty and development: into the 21st century.* T. Allen and A. Thomas (eds.) Oxford: Oxford University Press.

Toenniessen, G.H. (2002). 'Crop Genetic Improvement for Enhanced Human Nutrition.' *The Journal of nutrition* **132**(9 Supplement): 2943S-2946S.

Tripp, R. (2001). 'Can biotechnology reach the poor? The adequacy of information and seed delivery.' *Food Policy* **26**: 249-264.

Tuberosa, R. and S. Salvi (2006). 'Genomics-based approaches to improve drought tolerance of crops.' *Trends in Plant Science* **11**(8): 405-412.

Turner, M.R. (1994). 'Trends in India's seed sector.' Paper presented at Asian Seed '94, Chiang Mai, Thailand. 27-29 September.

Uijtewaal, B. (2006). 'Development of sustainable control of diamondback moth in cabbage and cauliflower by public private partnership.' In: *Science and technology policy for development; dialogues at the interface.* L. Box and R. Engelhard (eds.) London: Anthem Press.

United Nations (1992). 'The Convention on Biological Diversity, Article 2 (Use of terms).'

United Nations (2007). 'The Millennium Development Goals Report.' New York: United Nations.

Vaeck, M., A. Reynaerts, H. Höfte, S. Jansens, M. De Beuckeleer, C. Dean, M. Zabeau, M. Van Montagu and Jan Leemans (1987). 'Transgenic plants protected from insect attack.' *Nature* **328**: 33-37.

Van den Belt, H. and A. Rip (1990). 'The Nelson-Winter-Dosi model and synthetic dye industry.' In: *The Social Construction of Technological Systems*. W.E. Bijker, T. Hughes and T.J. Pinch (eds.) Cambridge, Massachusetts: MIT Press.

Van der Ploeg, J.D. (1993). 'Potatoes and Knowledge.' In: *An anthropological critique of development.* M. Hobart (ed.) London: Routledge: 209-227.

Van der Ploeg, J.D. [1999] (2003). *The Virtual Farmer*. Assen: Royal van Gorcum.

Van der Ploeg, J.D. (2008). *The new peasantries: struggles for autonomy and sustainability in an era of empire and globalization.* London: Earthscan.

Van der Ploeg, J.D. and A. Long (1994). 'Endogenous Development: Practices and perspectives.' In: *Born from within. Practices and perspectives of endogenous rural development.* J.D. van der Ploeg and A. Long (eds.) Assen: Van Gorcum.

Verbeure, B., E. van Zimmeren, G. Matthijs and G. Van Overwalle (2006). 'Patent pools and diagnostic testing.' *Trends Biotechnol* **24**(3): 115-20. Mar.

Verhoog, H., M. Matze, E.T. Lammerts van Bueren and T. Baars (2003). 'The role of the concept of the natural (naturalness) in organic farming.' *Journal of Agricultural & Environmental Ethics* **16**(1): 29-49.

Von Hippel, E. (2005). *Democratizing Innovation*. Cambridge, Massachusetts & London, England: The MIT Press.

Vroom, W. (2008a). 'Reconsidering agricultural modernization: three dimensions of questioning and redesigning biotechnologies for international agricultural development.' In: *Reconstructing biotechnologies: critical social analyses*. G. Ruivenkamp, S. Hisano and J. Jongerden (eds.) Wageningen: Wageningen Academic Publishers: 143-164.

Vroom, W. (2008b). 'Redesigning biotechnology: experiences of a public private partnership in the development of pro-poor transgenic cabbages in India.' *The European Journal of Development Research* **20**(3): 398-414.

Vroom, W. (2009). 'Linking upstream genomics research with downstream development objectives – The challenge of the Generation Challenge Programme.' *Tailoring Biotechnologies* **4**(3).

Vroom, W., G. Ruivenkamp and J. Jongerden (2007). 'Articulating alternatives: biotechnology and genomics development within a critical constructivist framework.' *Graduate Journal of Social Sciences* **4**(1): 11-33.

Wang, S., D.R. Just and P. Pinstrup-Andersen (2006). 'Tarnishing Silver Bullets: Bt Technology Adoption, Bounded Rationality and the Outbreak of Secondary Pest Infestations in China.' American Agricultural Economics Association Annual Meeting, Long Beach, CA, July 22-26, 2006. Online available at: http://www.grain.org/research_files/SWang_tarnished.pdf (last accessed 17 September 2008).

Watts, J., R. Mackay, D. Horton, A. Hall, B. Douthwaite, R. Chambers and A. Acosta (2003). 'Institutional learning and change: an introduction.' *ISNAR Discussion Paper No. 03-10, October 2003.*

Weatherford, J. (1988). *Indian givers: How the Indians of the Americas transformed the world*. New York: Crown Publishers.

Whatmore, S. (1995). 'From farming to agribusiness: the global agro-food system.' In: *Geographies of Global Change. Remapping The World In The Late Twentieth Century*. R. Johnston, P. Taylor and M. Watts (eds.) Oxford: Blackwell: 36-49.

Winner, L. (1985). 'Do Artifacts Have Politics?' In: *The social shaping of technology*. D. MacKenzie and J. Wajcman (eds.) Buckingham: Open University Press: 26-38. (Reprinted from 'Do Artifacts Have Politics' by Langdon Winner in Daedalus, 109, 121-36, 1980).

Witcombe, J.R., K.D. Joshi, R.B. Rana and D.S. Virk (2001). 'Increasing genetic diversity by participatory varietal selection in high potential production systems in Nepal and India.' *Euphytica* **122**(3): 575-588.

Woolgar, S. (1991). 'The Turn to Technology in Social Studies of Science.' *Science, Technology and Human Values* **16**(1): 20-50.

Woolgar, S. (1996). 'Technologies as cultural artefacts.' In: *Information and communication technologies. Visions and realities*. W.H. Dutton (ed.) Oxford: Oxford University Press.

Woolgar, S. and G. Cooper (1999). 'Do artefacts have ambivalence? Moses' bridges, Winner's bridges and other urban legends in S&TS.' *Social Studies of Science* **29**(3): 433-449. Jun.

Ye, X., S. Al-Babili, A. Klöti, J. Zhang, P. Lucca, P. Beyer and I. Potrykus (2000). 'Engineering the provitamin A (beta-carotene) biosynthetic pathway into (carotenoid-free) rice endosperm.' *Science* **287**(5451): 303-5. Jan 14.

Yin, R.K. (2003). *Case study research: Design and methods. Third Edition*. Thousand Oaks, London, New Delhi: SAGE Publications.

Young, N. (1990). *Seed potato systems in developed countries: Canada, the Netherlands and Great Britain*. Lima: International Potato Center.

Yu, J., S. Hu, J. Wang, G.K.-S. Wong, S. Li, B. Liu, Y. Deng, L. Dai, Y. Zhou, X. Zhang, M. Cao, J. Liu, J. Sun, J. Tang, Y. Chen, X. Huang, W. Lin, C. Ye, W. Tong, L. Cong, J. Geng, Y. Han, L. Li, W. Li, G. Hu, X. Huang, W. Li, J. Li, Z. Liu, L. Li, J. Liu, Q. Qi, J. Liu, L. Li, T. Li, X. Wang, H. Lu, T. Wu, M. Zhu, P. Ni, H. Han, W. Dong, X. Ren, X. Feng, P. Cui, X. Li, H. Wang, X. Xu, W. Zhai, Z. Xu, J. Zhang, S. He, J. Zhang, J. Xu, K. Zhang, X. Zheng, J. Dong, W. Zeng, L. Tao, J. Ye, J. Tan, X. Ren, X. Chen, J. He, D. Liu, W. Tian, C. Tian, H. Xia, Q. Bao, G. Li, H. Gao, T. Cao, J. Wang, W. Zhao, P. Li, W. Chen, X. Wang, Y. Zhang, J. Hu, J. Wang, S. Liu, J. Yang, G. Zhang, Y. Xiong, Z. Li, L. M.C. Zhou, Z. Zhu, R. Chen, B. Hao, W. Zheng, S. Chen, W. Guo, G. Li, S. Liu, M. Tao, J. Wang, L. Zhu, L. Yuan and H. Yang (2002). 'A draft sequence of the rice genome (Oryza sativa L. ssp. indica).' *Science* **296**(5565): 79-92. Apr 5.

Zannou, A. (2006). *Socio-economic, agronomic and molecular analysis of yam and cowpea diversity in the Guinea-Sudan transition zone of Benin.* Wageningen: PhD thesis, Wageningen University.

Zhao, J.Z., J. Cao, Y. Li, H.L. Collins, R.T Roush, E.D Earle and A.M Shelton (2003). 'Transgenic plants expressing two Bacillus thuringiensis toxins delay insect resistance evolution.' *Nature Biotechnology* **21**(12): 1493-1497. Dec.

Zimmerer, K.S. (1991). 'The regional biogeography of native potato cultivars in highland Peru.' *Journal of Biogeography*(18): 165-178.

Summary

Agriculture and food production are of ongoing high priority in international development debates, focused on the alleviation of extreme poverty and the eradication of hunger. One of the many ways of contributing to agricultural development for the resource poor is the development of new crop varieties that are more resistant to disease and pests, and that produce more in unfavourable circumstances such as drought or on poor soils. The contemporary revolutionary pace of innovation in genomics, marker assisted breeding and biotechnology creates a background in which there is a lot of scope for improving existing crop varieties, and addressing some of the most pressing problems that farmers are coping with. However, while the technical potential to improve crop varieties may be increasing, that does not automatically mean that we are actually able to solve problems in agricultural production. New technologies may never reach farmers, may be prohibitively expensive, or may solve only a very limited part of the problem that farmers are facing in practice. These observations have led to an interesting debate on how to make sure that the potential of modern genetic technologies can actually contribute to solving the problems of resource poor farmers in agricultural production.

In this debate it is commonly recognized that not any technology that is successful in a western production system, can simply be parachuted into a farming system in a developing country, and be expected to work. Instead, a notion of 'appropriate technology development' has become crucial in international development debates, drawing attention to the fact that technologies and development projects need to be adapted to the specific problems at hand, and the circumstances in which a technology has to work. However, this notion of 'appropriate technology' is far from being self-evident or straightforward, and no concrete recipe exists for defining what constitutes appropriate technology in a given set of circumstances. For that reason, the question emerges how this notion of 'appropriateness' is being operationalized by different project in practice.

This thesis engages with the debate on appropriate technology development by moving beyond a technical perspective on what constitutes 'appropriate technology' and by focusing on how different approaches to agro-technological development create different social roles for technology developers and farmers in innovation processes and production systems. This leads to a genealogy of strategies for agro-technological development in which farmers may be treated as 'recipients of technology', or may be involved as 'co-innovators', and in which technology developers may present themselves as 'solution providers' or 'service providers'. Insight in those different approaches can contribute to a clearer debate on the potential role of biotechnology in agricultural development and the reduction of poverty.

The first chapter of the thesis introduces the field of international agricultural development, and the role that genetic technologies can play in that context. The Green Revolution is introduced and discussed as an important previous experience in the large scale, planned modernisation of agriculture in the developing world. The question whether the Green Revolution was a success is still being answered in very different ways. This illustrates that the success of a technology can only be measured with respect to a specific perspective on agricultural development, and the question what exactly a new technology is for: an increase in productivity or the alleviation of poverty. The chapter also introduces the notion of reflexive development, which focuses on the learning dynamic in development projects, and the capacity to take comments, criticisms and concerns regarding agricultural development on board in innovation processes. Such reflexivity seems essential in order to answer the question what new technologies are for, and how it relates to a specific perspective on agricultural development. However, the way in which such reflexivity takes shape in practice will strongly depend on institutional and political factors. This is illustrated by a comparison of the historical backgrounds of the Green Revolution and the more recent 'Gene Revolution', which demonstrates that both processes are determined by entirely different dynamics, ideologies and (commercial) interests. This provides a starting point for this study, which investigates how contemporary projects use genetics and biotechnology for agricultural development, and tries to understand how technologies are made appropriate for resource poor farmers in developing countries.

Chapter 2 describes the research design of this study. It introduces the aim of the study, its main research questions, the three case studies and the methodology for data collection. In addition it reflects upon the suitability of the methodology to answer the research questions posed, and discusses the validity of the conclusions that are drawn based upon this explorative study.

Chapter 3 provides a more extensive conceptual background, deepening the discussion started in the first introductory chapter. Agricultural development is discussed in terms of modernisation and industrialisation processes which are argued to both contribute to a relatively homogeneous approach to agricultural development, and to the externalization of many aspects of farming practice like breeding and seed management. While this may have been a highly successful model of agricultural development in some parts of the world, its appropriateness for resource poor farmers is challenged. Agriculture in developing countries is often small-scale and characterised by a high degree of variability and localized adaptation. The question is whether an alternative model of agricultural development is possible in which genetic breeding technologies are used, but without necessarily externalizing agricultural innovation to specialized breeding institutes or companies. The expectation is that farmers in some areas are likely to require a more open-ended approach to agro-technological development in which they are empowered in their own on-farm experimentation with new crop varieties. The questions is what such an approach to innovation would look like, and whether elements of such an approach can be witnessed in the case studies. Finally, the question is raised what such an approach would mean in practice for the material design of new technologies or crop varieties.

Chapter 4, 5 and 6 present three case studies of projects in which plant breeding and genetic technologies are used to develop new crop varieties with interesting traits for resource poor farmers in developing countries.

Chapter 4 presents the case of the Collaboration on Insect Management for Brassicas in Asia and Africa (CIMBAA); a public private consortium in India which aims to develop a cabbage variety which is resistant against the diamondback moth. This insect is currently causing big losses in cabbage cultivation in India, and the CIMBAA consortium hopes to address this problem by engineering Bt insect resistance into a cabbage variety. The case study touches upon several dimension and aspects of making genetic technology appropriate for resource poor farmers, including the technical design of the gene construct that is used, the structuring role of intellectual property in the consortium, and the scope of stakeholder involvement in this project. The case is taken as a main illustration of how extensive efforts to reach resource poor farmers remain within the limits of an already existing industrial production system, in which the role of an external seed supplier is legitimized and consolidated. The innovation process in this case is characterised by its treatment of farmers as recipients of technology, and by their indirect representation in the project, rather than by their direct involvement.

Chapter 5 presents a set of initiatives of the International Potato Centre in Peru (CIP). The Peruvian Andes are the centre of origin of potato, and local traditional potato production is characterised by the use of a wide diversity of landraces. The use of modern improved potato varieties may boost productivity for farmers, but is feared to lead to the replacement of these native potato varieties, which are an important resource of genetic diversity for future plant breeding, as well as an important culinary and cultural resource for Andean potato farmers. For this reason, CIP is experimenting with participatory breeding programmes, the repatriation of native potato varieties, and the marketing of traditional potato varieties. These initiatives are argued to challenge the common bias in agricultural modernisation towards a narrowing genetic base, and the specialization on the cultivation of a very limited number of crop varieties. In addition, the centre is experimenting with virus resistance kits, which may significantly slow down the degradation of potatoes because of virus infestation. The combination of improved virus resistance of potatoes, diagnostic techniques and improved virus management practices may allow farmers to sustainably produce their own seed potatoes, providing them with a reasonable alternative to commercially available seed potatoes. The case study discusses the extent to which the technological interventions by CIP are capable of challenging ongoing trends towards an industrialisation of potato production, and capable of empowering farmers in their own on-farm seed potato production. Finally, the case represents an example of how farmers can be involved in agricultural development as co-innovators with specific valuable and complementary knowledge and expertise.

Chapter 6 presents the work of the Generation Challenge Programme (GCP), which is committed to the use of upstream comparative genomics research for the development of drought resistance traits in crops of interest to resource poor farmers. The chapter evaluates the priority setting exercise conducted by GCP and the way in which the programme tries to make sure that its research outputs are actually taken up by downstream research partners. The chapter discusses some of the potential difficulties in this process and explores the potential of 'complementary innovation systems' in order to meaningfully link upstream science-led genomics research and downstream bottom-up breeding programmes. The Genotyping Support Service (GSS) is expected to play an important role in that respect. The GSS is a very accessible service which allows the outsourcing of molecular analyses to specialized institutes for a variety of projects. This initiative is taken as a potentially very interesting approach to agro-technological development that shifts focus from the development of a technical solution, to the provision of a technical service. The GSS may as such constitute a technical interface between upstream genomics research and downstream variety development. The case also is a clear example of treating local research partners and farmers as co-innovators in agricultural development.

Chapter 7 brings together the analyses of the three case studies and evaluates how the different projects have practically operationalized the objective to develop 'appropriate technology' for the agricultural development of resource poor farmers. This leads to an extensive discussion on the different dimensions in which appropriateness of technological innovation is interpreted and reconsidered, and to the formulation of a contemporary understanding of what appropriateness means in practice. The multi-dimensional understanding of appropriateness that emerges from this analysis is taken as an argument for 'reflexive biotechnology development' as an approach to agro-technological innovation. The chapter further reflects upon the extent to which the material design of the various genetic technologies in the case studies is related to specific structures of production or innovation systems, and the extent to which the use of genetic technologies in plant breeding necessarily leads to an externalization of the innovation process. Some practical implications of the study for contemporary innovation policy are discussed and new questions for future research are formulated. This leads to the formulation of the following practical recommendations for policy makers or research managers in the field of agro-technology development:

- Encourage reflection upon the alignment between research approaches and the model of agricultural development that may be most appropriate for the intended beneficiaries, especially in terms of the production system and innovation system that is created or supported.
- Encourage the experimentation with breeding strategies that focus on the release of a wider set of new varieties and that aim to complement the cultivation of traditional varieties, rather than replace it. In addition, encourage the experimentation with crop varieties that allow farmers to be involved in their on farm seed production and hence may be instrumental in the strengthening of the informal seed sector. These strategies may not be

appropriate for all farmers in all regions, but they provide a potentially valuable alternative to the trends of an externalization of the means of production and the decrease of local innovation capacity, and may as such be highly appropriate for farmers in regions that have so far been left behind in attempts of agricultural development.

- Make optimal use of the complementarities between a genetic perspective on plant breeding and the capacities of farmers in variety selection, in order to maximize the relevance of new crop varieties.
- Invest in institutional learning processes and develop methods to measure research impacts along those lines.
- Encourage public private partnerships for the potential complementarity in capacities and the access to proprietary IP, but be careful in the kind of production system that is supported and the interests of the private partners to play a future role as technology provider. The model of agricultural modernisation that is interesting for the private partner may not always be the most appropriate model for resource poor farmers.
- Encourage stakeholder involvement and interactive research processes, but don't blindly extend this effort to upstream genomics research. In that domain, it makes much more sense to reflect on the way in which complementarities with different downstream partners can be exploited, and how new technologies and methodologies can be made available as a service, rather than as a concrete artefact.

About the author

Wietse Vroom was born in Tegelen, the Netherlands, on July 8th 1980 and grew up in Venlo. After an exchange year in Australia (1997-1998), he came to Wageningen to study Molecular Sciences. Although his initial interest was in biochemistry and microbiology, he developed an interest in social sciences during his studies and decided to do a second master thesis in Communication Sciences. In 2004 he graduated (with distinction) in Molecular Sciences at Wageningen University with a combined focus on Bacterial Genetics and Communication Sciences. The same year, he started his PhD research on the development of genetic technologies for resource poor farmers in developing countries, the result of which you are now holding. During his PhD research, Wietse has been involved in and chaired the organisation of several symposia and conferences on the social implications of genomics research and on intellectual property issues. His has given courses on 'Ethics in the Life Sciences', and has given interview trainings and focus group workshops. For his study, he has done research in India, Peru and Mexico. He is currently living and working in Guatemala with his wife Marleen.

www.ingramcontent.com/pod-product-compliance
Lightning Source LLC
Chambersburg PA
CBHW061210240926
51846CB00006B/1169